Science for
Physical Geographers

Science for Physical Geographers

Donald A Davidson

A HALSTEAD PRESS BOOK

John Wiley & Sons
New York

© Donald A. Davidson 1978

First published 1978 by Edward Arnold (Publishers) Limited
London

Published in the U.S.A
by Halsted Press, a Division of
John Wiley & Sons, Inc., New York.

Davidson, Donald A
    Science for physical geographers.

    "A Halstead Press book."
    Bibliography: p.
    Includes index.
    1. Science. 2. Physical geography. I. Title.
    Q161.2.D38   500.2   78–11957

ISBN 0–470–26556–6

Printed in Great Britain

Acknowledgements

The author and publishers wish to thank the following for permission to reproduce or modify copyright material:

Addison Wesley Publishing Co. Inc. for figures 3.15, 3.16 © 1963 and figures 6.1, 6.4 © 1974;

Allen and Unwin Ltd., for figure 2.1;

The British Ceramic Society for figures 7.6, 7.7, 7.8, 7.9 and 7.10;

Cambridge University Press for figures 1.1 and 7.12;

Collier-Macmillan International Inc. for figures 4.5, 4.10, 4.11 and 7.11;

J. R. H. Coutts for figure 6.8;

The Literary Executor of the late Sir Ronald A. Fisher, FRS, Dr Frank Yates, FRS and Longman Group Ltd., London, for permission to reproduce the table of t values from their book *Statistical Tables for Biological, Agricultural and Medical Research* (6th edition, 1974) Appendix 5;

W. H. Freeman and Co. for figure 7.4;

Harper and Row Inc., for figures 2.3, 2.4, 2.6 and 4.7;

D. C. Heath and Co., for figures 6.5, 6.6, 7.1, 7.21 and 7.22;

HMSO, Edinburgh for figure 6.8;

Hutchinson and Co. Ltd., for figure 7.24;

The International Union for Pure and Applied Chemistry for the periodic table, Appendix 3;

McGraw-Hill Inc., for figures 7.5, 8.1 and for Appendix 4;

Macmillan Publishing Co. Inc., for figures 4.5 and 4.10;

Methuen and Co., Ltd., for figure 6.2;

The Open University Press for figure 4.3;

Oxford University Press for figure 8.5;

Pion Ltd., for figure 7.18;

Prentice-Hall Inc., for figures 4.1 and 8.4;

W. B. Saunders Co., and Dr P. Odum for figure 4.8;

Spon Ltd., for figure 7.14

Contents

Preface

This book resulted from an awareness that many students of physical geography lack much formal training in science. Its aim is to present a concise integrated account of the fundamentals of chemistry and physics relevant to physical geography, so that students can aquire a basic knowledge of science. The logic of the approach and the assumed level of science background are described in the introduction.

My initial intention was to write with Mr Graham Sumner a book which included introductory mathematics as well as physics and chemistry. It soon proved too ambitious to include all of this in one book of reasonable length and thus two have resulted, this one and the companion text by Graham Sumner called *Mathematics for Physical Geographers*.

The objective and general approach of this book were discussed at length with Graham Sumner and I wish to record his friendly and constructive collaboration. Over the last three years I have received from various people help and encouragement which I wish to acknowledge. During 1974 I spent much of the year working in Carleton University, Ottawa where I was very much influenced by Professor Peter Williams whose own work on the physics of frozen soil is a clear illustration of the applicability of science to physical geography. In addition, Professor Williams has given advice on various chapters of the book. Thanks are also extended to Mrs Catriona Gardner, Dr Chris Park and Dr Keith Smith for constructive suggestions. James Duff, Karen Eide, Barry Keegan, Graham Kerr, Mary-Anne Piggott and Paul Reynolds, all students at Strathclyde University, gave comments on a final draft. Last, but by no means least, the author acknowledges with affection the encouragement and support given to him by his wife, Caroline. Despite all the assistance, I wish to stress that I alone am responsible for any final misrepresentations or errors in the book.

Lenzie
September, 1977.   Donald A. Davidson.

*

Introduction

An outstanding development in physical geography over the last twenty years has been an increasing emphasis upon scientific forms of analysis. This is being achieved largely by the application of statistical techniques within the whole spectrum of teaching and research so that numeracy is now an essential ability for physical geographers. The use of such methods requires a rigorous knowledge of scientific method. Statistical analysis is only part of scientific method and the trend of increasing specialization means that physical geographers must continue to develop expertise in such disciplines as soil science, ecology, meteorology, hydrology: this is only possible if they have had some training in the fundamentals of science and scientific investigation. For example, physical geographers must be competent to design and execute experiments. With a firm scientific training they can make use of methods and theories developed in other branches of science and at the same time play an integral role in the advancement of earth or atmospheric science. The initial emphasis on description and classification in the earth and biological sciences has now been superseded by the use of analytical techniques in order to reveal the structures and functional mechanisms of systems on a spatial or temporal basis. For example, in soil science where so much effort has been and still is being spent on soil classification, increasing consideration is being given to soil processes. Such an approach requires a great deal of field and laboratory experimentation. Just as it is wrong in statistics to use a technique without knowledge of its assumptions and limitations, so in soil geography students must understand the basic principles of the techniques which they use. This is only possible if their physics and chemistry are of a reasonable standard. Exactly the same type of argument should be put forward for all subjects related to physical geography. Clearly if physical geographers are going to keep pace let alone make contributions to these subjects, then it is essential for them to have a more comprehensive foundation in science than has normally been the case in the past.

The need in physical geography for a basic understanding of physics and chemistry requires little emphasis. The rigorous study of processes within the physical environment is only possible if the relevant fundamental concepts in physics and chemistry are understood. For example, geomorphologists concerned with stream flow have to develop an expertise in fluid mechanics. The development of theory means that solutions have to be derived from first principles and mathematics is integral to such an approach. In models of hillslope development, for example, advanced concepts in calculus are involved as well as an intimate knowledge of hillslope forces and soil mechanics. The text *Theoretical Geomorphology* by Scheidegger (1970) is only intelligible if the reader has an extensive command of mathematics and physics.

Another trend in physical geography is the increasing emphasis being placed on the applied aspects of the subject. As physical geographers develop expertise in specific topics, so they can apply their techniques and knowledge to practical problems. The resultant answers must be obtained by rigorous scientific analysis and must display a level of scholarship similar to other applied sciences. Physical geographers must therefore be able to understand and employ the fundamentals of pure science if their work is to be respected within the scientific community and if their results are to be of practical use.

The aim of this book is to introduce fundamental principles of physics and chemistry, selecting only those concepts of relevance to physical geography. Necessarily there are a large number of subject areas which are little more than touched upon in this book. Clearly it is impossible to present in one book all the physics and chemistry which may be encountered in physical geography. Although a serious attempt is made to select scientific principles of relevance to most of physical geography, it is inevitable that a final bias may reflect the author's particular interest in soil science.

An encyclopedic approach to physics and chemistry was considered, but was not adopted. Instead it seemed more important to approach the topic in an integrated manner, the aim being for readers to obtain a broad comprehension of physics and chemistry. A dominant theme in physical geography is the analysis of processes and how these processes interact to produce characteristic forms. Integral to such an approach is a knowledge of the physical and chemical nature of matter as well as comprehension of the energy laws which explain the nature and rates of processes. Indeed the study of physical environment on the basis of energy defined systems, is now common, an approach well illustrated by Strahler and Strahler (1973). The focus on processes and energy in physical geography is reflected in the emphasis and recurring themes of this book. Thus the reader who picks up this book with the hope of finding out how to carry out a particular titration, will be disappointed. Instead it is felt more important for students of physical geography, like all environmental scientists, to have a broad firm grasp of basic chemical and physical principles. Upon such a foundation they can build more specific scientific expertise relevant to particular subject areas of physical geography.

The text is designed for students in first and second year university or equivalent courses though it should also be useful for advanced courses; the assumption throughout the book is that the reader has had little formal training in physics and chemistry—nothing beyond 'O' level. For the reader with such training, it is hoped that the book will demonstrate the relevance of physics and chemistry to physical geography as well as to other environmental sciences. A knowledge of basic mathematics is assumed for this book; readers without such skills are referred to the companion volume, *Mathematics for Physical Geographers* by G. N. Sumner.

1

Dimensions, scales and units

The aim of any scientific investigation is the provision of an answer to a problem. In the first instance the investigator encounters some unexpected fact, phenomenon or pattern and this motivates him to design suitable experiments. The nature and specification of the problem will condition the general type of approach to this analysis. Appropriate evidence is collected and processed to allow the investigator to try to solve the problem. In other words, by the production of an explanation, the unexpected is transformed into the expected. In essence then, scientific enquiry involves problem identification, research design, data collection and statistical analysis, and interpretation in order to provide an explanation of the problem. The major underlying assumption throughout this book is that a basic knowledge of physics and chemistry not only helps physical geographers to account for phenomena in the physical environment, but also helps them to identify and investigate problems.

The training of a physical geographer, like any other scientist, ought to include the foundations of physics and chemistry, as well as of mathematics, statistics and the philosophy of science. For texts which use geographical examples, the reader is referred to Sumner (1978) and Wilson and Kirkby (1975) for mathematics, to Gregory (1973), Mather (1976), Hammond and McCullagh (1974) for statistics and to Harvey (1969) for a detailed analysis of philosophical issues. Mosley and Zimpfer (1976) provide a succinct discussion of the types of explanation used in geomorphology. Computing is increasingly being viewed as an indispensable skill and an appropriate geographical text is Dawson and Unwin (1976). These texts all reflect the growing scientific nature of the subject and today, the student of physical geography has to obtain a breadth and depth of expertise unthought of in the past.

One integral component of scientific investigation is measurement which involves such themes as dimensions, scales, units, methods of measurement and assessment of errors. Given the importance of these topics with regard to physics and chemistry, a discussion of them seems an appropriate way to begin, though a detailed discussion of the presentation of results and assessment of errors is postponed until the last chapter.

1.1 Dimensions

Proper reference to units of measurement is an essential principle of scientific method. The aim of any experimental scientist is to search for order among measured variables and this is only possible if these variables have properly defined dimensions. For example, the measurement of land area involves dimensions of length by length (L^2) whilst stream discharge involves length by length by length per time unit ($L^3 T^{-1}$). Length and time are base dimensions and as indicated, are represented by L and T respectively. Another important base dimension is mass (M). Many other dimensions can be derived from these three base units, though as will be shown with the international system of measurement, it is usual to include more base dimensions such as temperature and electric current.

Any distance clearly has the dimension L, while area has L^2, and volume L^3. Stream density, obtained by dividing the total length of streams by drainage area has the length dimension L/L^2 or $1/L$ or L^{-1}. Velocity is distance travelled in a unit of time (LT^{-1}) whilst acceleration is a rate of change of velocity with time and has the dimension LT^{-2}. A force is defined in terms of moving a mass with a particular acceleration and hence has the dimension resultant from multiplying mass and acceleration (MLT^{-2}). Stress or pressure is force divided by area and the dimensions can be calculated as

$$\frac{\text{force}}{\text{area}} = \frac{MLT^{-2}}{L^2} = ML^{-1}T^{-2}$$

Work is expressed as the distance over which a force is exerted so that its dimensions are $MLT^{-2}.L$ or ML^2T^{-2}. Dimensionally work and energy are identical. For example, a body of mass m travelling at speed v has a kinetic energy of $\frac{1}{2}mv^2$—this is a measure of the amount of work which the body can perform in being brought to rest. The dimensions of $\frac{1}{2}mv^2$ are $M(LT^{-1})^2$ which equals ML^2T^{-2}.

An equation is a mathematical statement of an equality and must be dimensionally consistent. In other words the dimensions on both sides of the equation must balance. Consider a simple example:

$$\text{density} = \frac{\text{mass}}{\text{volume}}$$

Suppose it is required to calculate the mass of a block whose density and volume are known. Then the equation can be written as

$$\text{mass} = \text{volume} \times \text{density}$$

Is this dimensionally correct? The dimensions of volume × density are $L^3.ML^{-3}$ which is M and thus the equation is correct.

As well as checking the dimensional equivalence of equations, dimensional analysis can be used in certain situations to suggest the nature of physical relationships. This is often demonstrated by considering the factors which influence the period of swing, say t, of a pendulum. The length of the pendulum (l) as well as the value of g, the acceleration due to gravity could well be hypothesized as the controlling factors. Thus the following statement can be written

$$t = f(l,g)$$

meaning that the period of the swing (t) is a function of l and g. This equation can be rewritten as

$$t = kl^a g^b$$

where k, a and b are dimensionless constants. The dimensions of this equation are

$$T = [L]^a \left[\frac{L}{T^2} \right]^b \tag{1.1}$$

where T represents the time dimension and L the length dimension. Equation (1.1) can be re-expressed as

$$T = [L]^{a+b}[T]^{-2b}$$

In order to balance the dimensions on either side of the equation,

$$a + b = 0$$

and

$$-2b = 1$$

Thus

$$b = -\tfrac{1}{2} \text{ and } a = \tfrac{1}{2}$$

This suggests that the actual form of the relationship between t, l and g is

$$t = k \sqrt{\frac{l}{g}}$$

a result which must then be tested by experimental procedures.

Of particular use are *dimensionless values*. The best known and simplest is π, the ratio of the circumference to the diameter of a circle. A further example of relevance to physical geography is the Reynolds number (Re), used to indicate the nature of stream flow; high values of Re indicate flow in the turbulent range and low values, flow in the laminar range (Leopold, Wolman and Miller 1964). The nature of these flow conditions is described in chapter 7. In this respect flow in a stream will depend on the friction (F_r) which will be influenced by the hydraulic radius (R), velocity of flow (V), the density of fluid (ρ) and the viscosity of the fluid (η). The viscosity of a fluid can be viewed in simple terms as the internal friction between its component layers. The hydraulic radius is the cross sectional area of a stream channel divided by the wetted perimeter.

Thus F_r is a function of (R, V, ρ, η)

Dimensions:

R (area/length)	L
V (distance/time)	LT^{-1}
ρ (mass/volume)	ML^{-3}
η (force × distance)/(area × velocity)	$\dfrac{(MLT^{-2})(L)}{(L^2)(LT^{-1})} = ML^{-1}T^{-1}$

The problem is to recognize a grouping of these variables so that a dimensionless value is obtained, and it is clear that if R, V and ρ are multiplied, the dimensions are the same as η (Bragg 1974). Thus the ratio $RV\rho/\eta$ is dimensionless and was found by Reynolds to be of value in fluid dynamics. For rivers

$$Re = \frac{RV\rho}{\eta} \text{ where } Re \text{ is Reynolds number}$$

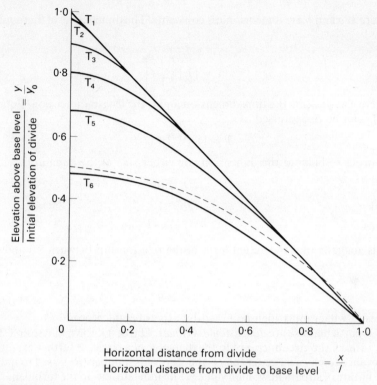

Fig. 1.1 Dimensionless graph (solid lines) showing slope development at times T_1 to T_6 by creep of an initially straight slope with a fixed divide (at $x = 0$) and a fixed base level (at $x = 1$). Broken line shows a steady state profile in equilibrium with slope-base downcutting at constant rate. (from Carson and Kirkby 1972, p. 298)

The use of dimensionless values is extremely helpful because comparisons can then be made easily. The index of shape, $1 \cdot 27 \, A/l^2$, where A is the area and l the longest axis of the area (Haggett *et al.* 1977) is another example of a dimensionless value. Dimensionless graphs are also useful since measurement scales have no effect on the form of the graphs. Graphs of slope development can be expressed in a dimensionless form (figure 1.1). The advantages of expressing a slope evolutionary model in this form are that the results are independent of measurement units and scale of analysis. This means for example, that measures from actual slopes of varying sizes can easily be compared when the results are expressed in dimensionless form.

1.2 Measurement scales

There are four scales of measurement and these depend upon the type of variables under analysis. A *nominal* scale is one of simple classification whereby individuals are assigned to particular classes, but no differences in rank or magnitude are implied between the classes. For example the numbers of different types of rock in an exposure of glacial till can be counted, but no order of importance is suggested in terms of rock type. With an *ordinal* scale it is possible to rank the classes, but magnitude of differences between classes is not specified. For example land can be classified into seven grades according to land use capability with class I the best and class VII the worst (Bibby and Mackney 1969). With an *interval* scale measurement is made on a continuous basis, but equal differences on the scale do not imply equal differences in magnitude. For example, temperature is measured in degrees Celsius, but 0 °C as the freezing point of water is *chosen* as zero on the scale. This means that 20 °C is not twice as hot as 10 °C. This would be the case with a *ratio* scale, for example with density. A soil with a density of $1\cdot5 \times 10^3$ kg m^{-3} is half as dense as a soil with a value of $3\cdot0 \times 10^3$ kg m^{-3}. With a density scale an absolute zero is automatically defined. In practice the distinction between interval and ratio scales may be academic, but the differences between these scales and nominal and ordinal ones are important. In an investigation, possible scales of measurement should be considered at the outset since the nature of statistical tests depends on the nature of the data. For example, if the data are in a nominal or ordinal form, then statistical analyses will be restricted to non-parametric techniques since they are based on comparison between distributions and not between parameters (Dixon and Massey 1957). This also illustrates the distinction between *continuous* and *discrete* observations. With nominal and ordinal scales discrete observations are made since numbers of occurrence are counted whilst with interval and ratio scales, continuous measurement is possible. These measurement considerations ought to be borne in mind at the experimental design stage given the implications for methods of statistical analysis and thus ultimate interpretation.

1.3 Units

The determination of physical quantities results in the production of numerical values which depend upon the units of measurement. For example, the discharge of a stream can be obtained by measuring the cross-sectional area of the channel and multiplying this by a measure of the average stream velocity. The result could be expressed as follows:

$$Q = 10\cdot31 \text{ m}^3 \text{ s}^{-1}$$

where Q represents discharge and the units are in terms of cubic metres per second. The reporting of the numerical value ought to take into account the errors associated with the determinations, a theme to be discussed in chapter 8.

The need for a standard internationally accepted set of units and abbreviations is obvious; several measurement systems are in use, but the one which is gaining wide acceptance is 'Le Système International d'Unités', usually referred to as the SI system (HMSO 1973). The core of this system is seven *base units* (table 1.1) from which are produced *derived units*. The base unit of mass is the kilogram; the international standard is a cylinder of platinum and iridium kept under controlled conditions by the International Bureau of Weights and Measures at Sèvres, near Paris. The other base units are listed in table 1.1. These units all have fundamental definitions which need not be considered here, but it is important to appreciate that all units used in science are derived from them. Some comments on temperature scales are appropriate. The Kelvin scale is an absolute one since no temperature is possible below zero on this scale. On the Kelvin scale water normally freezes at $273\cdot15$ K and boils at $373\cdot15$ K. With the Celsius scale the normal freezing

Table 1.1 SI base units.

Quantity	Name	Symbol
length	metre	m
mass	kilogram	kg
time	second	s
thermodynamic temperature	Kelvin	K
electric current	ampere	A
luminous intensity	candela	cd
amount of substance	mole	mol

point of water is set as $0\cdot00$ °C and the boiling point as $100\cdot00$ °C. The size of the degree on the Kelvin scale is the same as on the Celsius scale and in a strict sense should be called a Centigrade degree. Centigrade is popularly used as a scale, but the correct one to use is the Celsius or Kelvin scale. It is useful to note the distinction between 10 °C and 10 C°; the former refers to an actual temperature of 10 °C whilst the latter is a temperature difference, say between 20 °C and 30 °C.

Derived units are clearly obtained by the combination of base units and again the SI system specifies these derived units. A selected list of derived units is presented in appendix 1. All of these quantities will be explained in later parts of the book.

In the SI system certain conventions must be followed for quoting units and these are clearly specified by the Royal Society (1975). Symbols should not change in the plural and should not be followed by a full stop except when at the end of a sentence, e.g.

6 kg but not 6 kgs nor 6 kg.

The product of two units can be expressed as demonstrated by:

6 N m or 6 N.m

A quotient of two units can be expressed in the following ways:

6 m s^{-1}, 6 m.s^{-1} or 6 m/s

The dimensions in expressions which have more than two base units need not be separated by dots (see examples in appendix 1); indeed dots should only be used if there is any danger of misinterpretation. The solidus (/) should not be used more than once in an expression unless parentheses are used, e.g.

$W \text{ m}^{-1} \text{ K}^{-1}$ or W/(mK)

but not W/m/K

The SI system specifies a series of names and symbols to form decimal multiples of SI units (table 1.2). The obvious need for such multiples can be demonstrated by considering how the size of clay size particles ought to be expressed. Clay is usually taken to be individual particles smaller than $0\cdot002$ of a millimetre, i.e.

clay $< 0\cdot002$ mm

An alternative form of expression, again conforming to the SI system is as follows:

clay $< 2 \ \mu$m

Table 1.2 Names and symbols of multiples of SI units.

Factor	Prefix	Symbol	Factor	Prefix	Symbol
10^{12}	tera	T	10^{-1}	deci	d
10^{9}	giga	G	10^{-2}	centi	c
10^{6}	mega	M	10^{-3}	milli	m
10^{3}	kilo	k	10^{-6}	micro	μ
10^{2}	hecto	h	10^{-9}	nano	n
10^{1}	deca	da	10^{-12}	pico	p
			10^{-15}	femto	f
			10^{-18}	atto	a

Table 1.3 Units used with SI system. (Source: Royal Society 1975, p. 26.)

Physical quantity	Name of unit	Symbol for unit	Definition of unit
time	minute	min	60 s
time	hour	h	60 min
time	day	d	24 h
angle	degree	°	$(\pi/180)$ rad
angle	minute	′	$(\pi/10800)$ rad
angle	second	″	$(\pi/648000)$ rad
volume	litre	l	$10^{-3} m^3 = dm^3$
mass	tonne	t	10^3 kg = Mg
Celsius temperature	degree Celsius	°C	K

Micrometres are usually called microns. Note that no space is left between the prefix symbol and the unit symbol. The use of the multiples in table 1.2 avoids very large or small numbers in a result. A density measure of $1\cdot5 \times 10^3$ kg m^{-3} has already been mentioned, but clearly is a rather clumsy expression even though it is expressed in the SI units for density. However, adherence to the SI system would not be broken if the same result was reported as $1\cdot5$ g cm^{-3}.

Besides the units so far specified within the SI system, there are others presented in table 1.3 which are also accepted for use. Thus it is quite in order to discuss results in degrees Celsius. One time unit not specified in table 1.3, but which is also accepted in the system is the year (symbol a). Other names for particular multiples of SI units will be encountered, but it seems best to describe these as they arise in later chapters.

The need to adhere rigidly to the SI system requires no emphasis. It should also be noted that the term 'weight' does not appear in the SI system. The weight of a body can be obtained by multiplying its mass by g, the acceleration due to gravity, but if consideration is given to units it is clear that weight is a force (see chapter 3). Another misconception is that 'g' is a force of·gravity—instead it is an acceleration resulting from gravity. Thus not only is precision of expression achieved by dimensional considerations, but also clarity of thought. A useful habit is always to include units of variables in calculations. *Note:* Appendix 2 gives conversions of various units to the SI system.

2
The nature of matter

The nature, properties and structure of matter have always been the kernel of research in pure science. Over the centuries substances have been analysed into their simplest components called *elements*, which, by definition cannot be further subdivided. Prior to 1750 about 16 elements were known, but since then there has been a marked rise in their rate of discovery. Current chemistry text books list 105 elements, but three new super-heavy elements have recently been discovered and more are likely. Prior to the discovery of these, 92 elements were thought to exist naturally. The need for a standard international set of names and symbols for elements is obvious: the list as agreed by the International Union of Pure and Applied Chemistry is reproduced in table 2.1. Although a very large number of elements have been recognized, only a small number constitute the vast bulk of the earth's crust, oceans and atmosphere; in fact 10 elements comprise 99·2% of this zone and these are indicated in table 2.1. Examination of the figures in the table shows that 75·2% of the crust, oceans and atmosphere on a mass basis is made up of the two elements, oxygen and silicon, though hydrogen would be absolutely dominant if the calculation was done according to the number of atoms.

Elements nearly always occur in association with others, but exceptions are the noble gases (helium, neon, argon, krypton, xenon and radon). If a substance is made up of more than one element, then it can be either a mixture or a compound or some combination of these. In a mixture the individual elements are present in variable proportions. For example, a mixture of the gases oxygen and hydrogen could be made in any particular proportion. Compounds, on the other hand, have their elements present in particular and characteristic proportions; on the whole separation of elements in compounds is far more difficult than separation of elements in mixtures. Most substances in the physical environment are mixtures, not just of elements, but also of compounds. For example, the atmosphere is made up of a mixture of nitrogen, oxygen, argon, carbon dioxide and water vapour, plus several other gases in tiny quantities.

2.1 Introduction to atoms

For an understanding of the physical environment, ultimate recourse must be made to the nature of elements. The ancient Greeks first proposed that matter was composed of discrete particles, but it was not until the work of Dalton in the early nineteenth century that significant advances were made. Indivisible particles of elements were called *atoms* and combinations of atoms produced *molecules*. Dalton's *atomic theory* was an outstanding scientific achievement and his main postulates still remain central to modern atomic theory. Three of the postulates can be expressed as follows:

(1) An element is composed of extremely small particles called atoms. All the atoms in a given element are chemically identical.

(2) Atoms of different elements have different properties. In the course of an ordinary chemical reaction, no atom of one element disappears or is changed into an atom of another element.

(3) Compound substances are formed when atoms of more than one element combine. In a given pure compound the relative numbers of atoms of the elements present will be definite and constant.

(After Masterton and Slowinski 1973, p. 26.)

The latter two postulates have relevance to other general principles, in particular to the law of conservation of mass from postulate (2) and to the law of constant composition from postulate (3). These are topics to be discussed later.

The next major breakthrough was the realization that the atom was made up of component particles; the *electron* was the first to be identified in 1897 by Thomson. The mass of an electron (m_e) has been estimated to be $9 \cdot 11 \times 10^{-31}$ kg. It constitutes only a tiny fraction of the mass of atoms. Electrons are electrically charged particles and, by convention, are defined to have a unit negative charge. Since the discovery of the electron over 30 sub-atomic particles have been identified, but the three fundamental particles are the electrons, protons and neutrons. It was Rutherford who discovered that the atom has in its centre a *nucleus* which has a positive charge; electrons occur in various shells round the nucleus which contains the other two fundamental particles, protons and neutrons. Both these are about the same mass, but the proton has a unit positive charge, equal in magnitude but of opposite charge to the electron. In contrast, the neutron has no charge. Electrons are responsible for giving the cohesion between atoms as well as assigning a distinctive chemical characteristic to the atom; in contrast the particular role of neutrons and protons seems to be to give the atom mass. The *atomic number* (Z) of an atom is defined as the number of protons, which, in a neutral atom, will equal the number of electrons. The atomic number is thus an extremely important value in characterizing the nature of atoms. Atoms, all with the same atomic number, define one element and this is the reason why the elements in table 2.1 are listed according to their Z-values. The sum of neutrons and protons in an atom is called the *atomic mass number*. The nucleus in the carbon atom has 6 protons and 6 neutrons and the convention is therefore to label the carbon atom as having exactly 12 *atomic mass units* (amu). Dalton's postulates imply that individual elements have unique atomic structures. In a strict sense this only applies to the number of protons since different numbers of neutrons result in *isotopes* of the same element, the number of protons remaining constant. The atomic mass number (*A*) can be determined by adding the atomic number (Z) and the number of neutrons (*N*), i.e.

$$A = Z + N$$

The convention is to write the symbol for an element, for example, O for oxygen, with the atomic number as a subscript at the lower left of the element symbol and the atomic mass number as a superscript at the upper left. For

Table 2.1 List of elements according to atomic number. Note the British use of sulphur for S and aluminium for Al. The abundance of the 10 most common elements in the earth's crust, oceans and atmosphere is given (on a percentage mass basis). These elements account for 99·2% of this zone. (Abundance data from Masterton and Slowinski (1973)).

Atomic number	Symbol	Element name	Composition of earth's crust, oceans and atmosphere (%)
1	H	Hydrogen	0·87
2	He	Helium	
3	Li	Lithium	
4	Be	Beryllium	
5	B	Boron	
6	C	Carbon	
7	N	Nitrogen	
8	O	Oxygen	49·5
9	F	Fluorine	
10	Ne	Neon	
11	Na	Sodium	2·6
12	Mg	Magnesium	1·9
13	Al	Aluminium	7·5
14	Si	Silicon	25·7
15	P	Phosphorus	
16	S	Sulphur	
17	Cl	Chlorine	
18	Ar	Argon	
19	K	Potassium	2·4
20	Ca	Calcium	3·4
21	Sc	Scandium	
22	Ti	Titanium	0·58
23	V	Vanadium	
24	Cr	Chromium	
25	Mn	Manganese	
26	Fe	Iron	4·7
27	Co	Cobalt	
28	Ni	Nickel	
29	Cu	Copper	
30	Zn	Zinc	
31	Ga	Gallium	
32	Ge	Germanium	
33	As	Arsenic	
34	Se	Selenium	
35	Br	Bromine	
36	Kr	Krypton	

oxygen there are three isotopes and these are expressed as

$$^{16}_{8}O, \ 8 \ \text{protons}, \ 8 \ \text{neutrons}$$

$$^{17}_{8}O, \ 8 \ \text{protons}, \ 9 \ \text{neutrons}$$

$$^{18}_{8}O, \ 8 \ \text{protons}, \ 10 \ \text{neutrons}$$

In English these are referred to as 'oxygen 16, oxygen 17 and oxygen 18'. Some isotopes have particular names, such as $^{2}_{1}H$ (deuterium) and $^{3}_{1}H$ (tritium). In the physical environment it is common for one element to occur as a mixture of several isotopes. Sodium only occurs in one form ($^{23}_{11}Na$), but calcium, which occurs in chalk, limestone and bone is a mixture of $^{40}_{20}Ca$, $^{42}_{20}Ca$, $^{43}_{20}Ca$, $^{44}_{20}Ca$,

Table 2.1 *Cont.*

Atomic number	Symbol	Element name	Atomic number	Symbol	Element name
37	Rb	Rubidium	73	Ta	Tantalum
38	Sr	Strontium	74	W	Tungsten
39	Y	Yttrium	75	Re	Rhenium
40	Zr	Zirconium	76	Os	Osmium
41	Nb	Niobium	77	Ir	Iridium
42	Mo	Molybdenum	78	Pt	Platinum
43	Tc	Technetium	79	Au	Gold
44	Ru	Ruthenium	80	Hg	Mercury
45	Rh	Rhodium	81	Tl	Thallium
46	Pd	Palladium	82	Pb	Lead
47	Ag	Silver	83	Bi	Bismuth
48	Cd	Cadmium	84	Po	Polonium
49	In	Indium	85	At	Astatine
50	Sn	Tin	86	Rn	Radon
51	Sb	Antimony	87	Fr	Francium
52	Te	Tellurium	88	Ra	Radium
53	I	Iodine	89	Ac	Actinium
54	Xe	Xenon	90	Th	Thorium
55	Cs	Cesium	91	Pa	Protoactinium
56	Ba	Barium	92	U	Uranium
57	La	Lanthanum	93	Np	Neptunium
58	Ce	Cerium	94	Pu	Plutonium
59	Pr	Praseodymium	95	Am	Americium
60	Nd	Neodymium	96	Cm	Curium
61	Pm	Promethium	97	Bk	Berkelium
62	Sm	Samarium	98	Cf	Californium
63	Eu	Europium	99	Es	Einsteinium
64	Gd	Gadolinium	100	Fm	Fermium
65	Tb	Terbium	101	Md	Mendelevium
66	Dy	Dysprosium	102	No	Nobelium
67	Ho	Holmium	103	Lr	Lawrencium
68	Er	Erbium	104		Kurchatovium
69	Tm	Thulium			(proposed)
70	Yb	Ytterbium	105	Ha	Hahnium
71	Lu	Lutetium			(proposed)
72	Hf	Hafnium	106		(un-named)

$^{46}_{20}$Ca and $^{48}_{20}$Ca. Other isotopes of calcium can be created in a laboratory, but in contrast to those specified above, are *unstable*.

Atoms which decompose are *radioactive* and the result is the emission from the nucleus of an excess of energy. Such emission can take three forms:

(1) *alpha activity*: the expulsion of two protons and two neutrons which is a helium nucleus or alpha particle ($^4_2$He);

(2) *beta activity*: the expulsion of electrons;

(3) *gamma activity*: the emission of ultra-short-wave radiation. (For an explanation of radiation, see section 2.3 later in this chapter.)

Thus an atom undergoing radioactive decay can lose sub-atomic particles by alpha and beta activity as well as an additional loss of energy by gamma activity. The relevance of this is that the physical environment is composed of isotopes which are predominantly stable and this is reflected in the abundance of certain elements in contrast to others (table 2.1). The most abundant elements are those with the most stable nuclei, a characteristic which has been related to the number of neutrons compared to protons (Anderson, Ford and Kennedy 1973, pp. 12–13). A useful value is obtained by dividing the number of neutrons by the number of protons. The most stable value is about 1·0 for elements up to atomic number 20; for higher elements the stablest value increases slowly to about 1·6. The most stable isotope of lead is $^{208}_{82}Pb$ (126 neutrons and 82 protons) to give a stability ratio of 1·54.

Radioactive decay implies a change of one substance to another and the rate of decay is expressed as the *half-life*. The half-life of a radioactive isotope is the most likely time that will be required for half the atoms to be transmitted by radioactive decay into other substances. All radioactive substances have half-lives and these vary enormously according to the isotope. Suppose 10 g of $^{11}_{6}C$ was produced; after only 20 minutes 5 g of $^{11}_{6}C$ would be left because the half-life of $^{11}_{6}C$ is 20 minutes. 5 g of $^{11}_{6}C$ would have converted into $^{11}_{5}B$. After another 20 minutes another 2 ·5 g of $^{11}_{6}C$ would have changed to $^{11}_{5}B$ and so on. Thus a negative exponential curve is characteristic of radioactive decay. A most important process of radioactive decay is that of another carbon isotope, viz. $^{14}_{6}C$, since this decay is the basis of *radiocarbon dating* which is extensively used in Quaternary and archaeological research. In the earth's upper atmosphere radioactive $^{14}_{6}C$ is produced by the bombardment of $^{14}_{7}N$ by neutrons from outer space. In this process, each atom of $^{14}_{7}N$ absorbs a neutron and emits a proton to thus form $^{14}_{6}C$; this can be explained by noting that the atomic mass number remains unchanged at 14 whilst the atomic number drops from 7 to 6 which means a change from the element nitrogen to carbon. $^{14}_{6}C$ combines with oxygen to form carbon dioxide, which through atmospheric circulation and diffusion reaches the surface of the earth. The assumption has been made that the rate of production of $^{14}_{6}C$ in the upper atmosphere has been constant over recent earth history, but this is now open to question. For this reason radiocarbon dates have to be carefully calibrated before a calendar date is proposed. In any case this $^{14}_{6}C$ becomes available to living matter by being incorporated into atmospheric carbon dioxide which has a ratio of atoms of $^{14}_{6}C:^{12}_{6}C$ of the order of $1:10^{12}$. Plants absorb carbon in the same proportion by photosynthetic processes, but when the plant dies, the intake of $^{14}_{6}C$ stops, and this begins to decay to $^{14}_{7}N$ to cause a decreasing ratio of $^{14}_{6}C:^{14}_{7}N$. Radiocarbon data as reported in the journal *Radiocarbon* are based on the Libby half-life of 5 570 years. Suppose the problem is to date a buried organic horizon. In a radiocarbon laboratory the number of particles emitted from the sample in unit time can be measured and compared with the normal 13·5 counts per minute per gram for fresh organic matter. If, for example, a count of 7 per minute per gram was obtained from the sample, this would imply that half of the $^{14}_{6}C$ had been transformed and thus the age of the buried horizon corresponded to the half-life of $^{14}_{6}C$ in radiocarbon years. It should be stressed that the technique of radiocarbon dating is far more complicated than this simple introduction may imply. However, it can be appreciated that an elementary knowledge of atomic structure aids an understanding of such an important technique as radiocarbon dating. Other isotopes can also be used for dating rocks using radiometric techniques; the best known are isotopes of potassium, uranium, rubidium, thorium and lead. The problem has been the gap in dating material in the time range between the radioactive isotope of carbon and these other isotopes. For example radiocarbon dating is possible up to around 60 000 B.P., whilst the potassium–argon method can cover almost the entire geological time scale, though the dating of rocks younger than 100 000 years by this method is very difficult. All these other methods of radiometric dating are in principle exactly analogous to the radiocarbon technique.

2.2 Molecules, ions and compounds

As already stated it is very rare in nature for substances to occur as single atoms, the exceptions being the noble or inert gases. Instead most substances owe their nature to the combination of atoms, and the smallest particle which can then exist and have the properties of the substance is the molecule. In a strict sense one atom of an inert gas is a molecule, but most molecules are made up of more than one atom and these can be of the same or different elements. Clearly a molecule is held together by bonding forces and these are far stronger than the forces between molecules. Consider the simple example of water; a molecule of water results from the bonding of two atoms of hydrogen and one atom of oxygen and can be expressed as

$$H—O—H$$

while the gas, oxygen is composed of molecules made up of two atoms of oxygen. The nature of these bondings forces is of extreme importance in understanding the nature of substances as well as chemical processes. A full description of bonding is postponed until chemical processes are discussed in chapter 5.

Not all substances are made up of individual molecules; instead many atoms are linked in a three-dimensional lattice. In such a structure the atoms can have electrical charges—either positive or negative—and are called *ions*. If an electron is removed from a neutral atom, then an ion with a positive charge is produced and this is called a *cation*. An electron can also be added to a neutral atom to produce an atom with a negative charge—an *anion*. A sodium cation results from the loss of an electron from a neutral Na atom and is represented as Na^+; similarly a chlorine anion (Cl^-) can be produced. Common table salt, or sodium chloride, is composed of a lattice of sodium and chlorine ions (figure 2.1). Thus sodium chloride does not occur as a collection of individual molecules, but is an example of a *lattice compound* since each sodium cation is linked to 6 chlorine anions and vice versa. The nature of crystals is related to their lattice structure, a theme to be developed in chapter 7.

Electrons encircle nuclei and are kept in particular shells because of a complex set of forces within the atoms. The removal of an electron from a neutral atom demands the expenditure of energy. In particular the energy which is necessary to remove an electron from an isolated gaseous atom is called the *ionization energy*; removal of the first electron will require less energy than the second and so on, which means that there is a first ionization energy, second ionization energy, etc. The parallel situation is the gain in energy, called the *electron affinity*, when an electron is introduced to a gaseous atom.

The bonding of atoms to form compounds can only be understood if some other properties of atomic structure are considered. The relative masses of

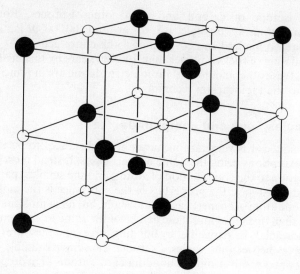

Fig. 2.1 Schematic diagram of a part of a sodium chloride crystal. The ions have been represented by small symbols for clarity. In fact they are large enough to touch each other. Full circles, Cl; open circles, Na. (from Stamper and Stamper 1971, p. 14)

individual atoms has been seen as a topic of fundamental importance since Dalton's pioneering work. Clearly it is impossible to determine directly the mass of an atom—the smallest mass which can be weighed contains of the order of 10^{17} atoms (Anderson, Ford and Kennedy 1973). Instead some relative scale has been produced by comparing the mass of individual atoms to one particular atom. The hydrogen atom would seem the obvious standard to accept since it is the lightest atom, but experimental problems arise with hydrogen when it is being used in a quantitative sense. For long oxygen was taken as the standard, but since 1961, carbon 12 ($^{12}_{6}C$) has been defined as the base unit with the mass of one atom of this isotope being given as 12 atomic mass units (amu). The masses of all other atoms can thus be expressed in relation to this value. Some examples are as follows:

Element	Isotope	Mass of isotope (amu)	Relative abundance (%)
Carbon	$^{12}_{6}C$	12·000 (standard)	98·89
	$^{13}_{6}C$	13·003	1·11
Nitrogen	$^{14}_{7}N$	14·003	99·63
	$^{15}_{7}N$	15·000	0·37
Oxygen	$^{16}_{8}O$	15·995	99·76
	$^{17}_{8}O$	16·999	0·04
	$^{18}_{8}O$	17·999	0·20

(From Gymer 1973, p. 17.)

It can thus be appreciated that values for the masses of individual isotopes are in a strict sense dimensionless. In nature, elements occur as a mixture of isotopes, as demonstrated in the above figures, so that some average measure of mass is required for elements. Imagine 10 000 atoms of naturally occurring oxygen; 9976 of these would weigh 15·995 amu each, 4 would weigh 16·999 amu each and the final 20 would weigh 17·999 amu each. Thus the *average* mass of one atom of naturally occurring oxygen can be determined as follows:

$$(9976 \times 15\cdot995) + (4 \times 16\cdot999) + (20 \times 17\cdot999)/10\,000 = 15\cdot999 \text{ amu}$$

This value is called the *atomic weight* for oxygen. Atomic weights are widely used in chemical calculations and the values for all the elements are given in the Periodic Table (Appendix 3). Comment must be made on the use of the term 'weight' given the statement in chapter 1 that this term is not included in the SI system. Atomic weight is widely used and accepted in chemistry, but values are dimensionless since they are obtained from a ratio of masses. This has led to a dimensional form called the *gram atomic weight* which is obviously expressed in grams. One gram atomic weight is a particular mass for each element—the numerical value is the same as the atomic weight. Thus one gram atomic weight of oxygen has a mass of 15·999 g; two gram atomic weights would be 31·998 g etc. One gram atomic weight of oxygen will be obtained from a certain number of oxygen atoms, a number which is of particular importance.

The arbitrary scale of atomic mass units (amu) has been described and 1 amu has a mass of $1\cdot66 \times 10^{-24}$ g. One gram atom of oxygen is composed of 15·999 amu and thus the mass of one oxygen atom is $(15\cdot999 \times 1\cdot66 \times 10^{-24})$ g. If one gram atomic weight has a mass of 15·999 g and the mass of one oxygen atom is $(15\cdot999 \times 1\cdot66 \times 10^{-24})$ g, then division of the former figure by the latter gives the number of atoms; the result is $6\cdot022 \times 10^{23}$ and is called *Avogadro's number* (N). In this division, the same result is obtained for any element since the numerical value of the atomic weight would always appear in the numerator and denominator.

The gram atomic weight of any element is the mass of $6\cdot022 \times 10^{23}$ atoms; if the mass of the same number of molecules is established, then the result is the *gram molecular weight*. Consider common salt, sodium chloride; the gram molecular weight of this compound is the mass of $6\cdot022 \times 10^{23}$ molecules of sodium chloride. In a strict sense individual molecules of sodium chloride do not exist in the solid state since all the atoms of sodium and chlorine are linked in a cubic structure (chapter 7). In order to obtain the gram molecular weight of sodium chloride excluding this complication, the gram atomic weights of sodium (22·99 g) and chlorine (35·45 g) require to be added to give an answer of 58·44 g. If atomic weights are added instead of gram atomic weights, then the result is the *molecular weight*, the same as gram molecular weight except without the mass dimension. Say a molecule is formed by 2 atoms of X with 1 of Y; the molecular weight is obtained by multiplying the atomic weight of X by 2 and adding it to the atomic weight of Y. A molecule of water is composed of two atoms of hydrogen (atomic weight 1·01) and one of oxygen (atomic weight 16·00) and thus the molecular weight of water is 18·02.

It is cumbersome to have to refer to $6\cdot022 \times 10^{23}$ atoms or molecules; instead the term *mole* (abbreviation mol) is used to mean this precise quantity of substance. Thus 1·00 mol of carbon atoms has a mass of 12·00 g and also corresponds to 1 gram atom. The gram molecular weight of any substance is also identical in quantity to 1 mole. The quantity, a mole, is particularly useful for describing the concentration of solutions (chapter 7). If the substance is a gas, then the volume occupied by 1 mole depends upon the temperature and pressure. If the gas has a pressure of 101 325 Pa (one atmosphere), and if the temperature is 273·15 K, then 1 mol of the gas occupies $2·2414 \times 10^{-2}$ m^3; this quantity is known as the molar volume and reference will be made to it again when dealing with the gas laws (chapter 7).

2.3 Atomic structure, the periodic table and chemical formulae

One of Dalton's postulates states that a given pure compound will always be composed of the constituent atoms in the same proportions. This is expressed quantitatively in the formulae of compounds. A molecule of water (H_2O) results from the bonding of two hydrogen atoms with one oxygen. In a strict sense this is an *empirical formula* since the formula indicates the simplest whole number proportions of the constituent elements. In contrast, a *molecular formula* gives the actual numbers of atoms in a molecule—either the molecular formula is the same as the empirical formula (as is the case with H_2O), or is some multiple of it. The example often quoted is that of hydrogen peroxide (H_2O_2) which is used in the pre-treatment of soil samples prior to particle size analysis in order to remove the organic matter. The empirical formula for hydrogen peroxide is HO, but in practice individual molecules are made up of two atoms of hydrogen linked to two atoms of oxygen—hence the molecular formula of H_2O_2. Molecular formulae are used whenever possible, though their use in the case of non-molecular substances, such as NaCl, `is inappropriate.

Two methods are used to establish chemical formulae, either by a quantitative analysis of the compound or by predicting on theoretical grounds what the formula ought to be, assuming knowledge of the constituent elements. The first strategy is clearly necessary when little is known about the compound. It is the role of the analytical chemist to analyse such substances and the results are used to establish the empirical formula. Suppose 100 g of a mystery substance was discovered to consist of 26·58 g of potassium, 35·35 g of chromium and 38·07 g of oxygen, how is the formula calculated? It is important to stress that the formula does *not* consist of these three elements with their proportions indicated by their relative masses. Instead the task is to compute the number of atoms of the elements which are linked to constitute one molecule. In effect what has to be done is to calculate the number of gram atomic weights for each of the three elements since these will also indicate the relative numbers of atoms. This is achieved by dividing the three laboratory determined weights by the gram atomic weight of the appropriate element. The calculations are as follows:

$$\text{Number of gram atomic weights potassium} = \frac{26·58}{39·09} = 0·68$$

$$\text{Number of gram atomic weights chromium} = \frac{35·35}{52·00} = 0·68$$

$$\text{Number of gram atomic weights oxygen} = \frac{38·07}{16·00} = 2·38$$

Thus potassium, chromium and oxygen atoms occur in the proportion 0·68 : 0·68 : 2·38 and to derive the simplest formula these numbers are divided by the smallest, viz. to give the ratio $1 : 1 : 3 \cdot 5$ which must be multiplied by 2 to give whole numbers, i.e. $2 : 2 : 7$. In other words the empirical formula for the mystery substance is $K_2Cr_2O_7$, potassium dichromate. It is very unlikely for a physical geographer to have to perform such an analysis; the above example has been included to illustrate the quantitative nature of a chemical formula since such an appreciation is essential for an understanding of chemical reactions, for example, weathering processes. For instance a physical geographer may require the formula for calcium carbonate, but there is no need for a quantitative analysis of calcium carbonate in order to establish the empirical formula $CaCO_3$; instead it is possible to postulate this formula after a few moments if more is known about the nature of these atoms. Attention must therefore return to further consideration of atomic structure and this has the added advantage that the principles of electromagnetic radiation can also be described.

It will be recalled that an atom consists of a nucleus containing neutrons and protons and the nucleus is encircled by electrons moving in different shells. As already suggested the electronic structure of atoms conditions many chemical properties—in particular the bonding of atoms depends upon the electrons in the outermost shell. It is thus understandable why there has been so much research into electronic structure over recent decades. The striking feature about an atom is that the vast bulk of its volume is empty space, an important result obtained by Rutherford. The problem is to understand the behaviour of electrons in all this space especially when it is realized that the principles of classical mechanics are inapplicable. The behaviour of electrons encircling a nucleus cannot be explained by analogy with billiard balls flying round some distant nucleus. Classical mechanics was derived to explain the patterns of movement of large bodies; instead *quantum mechanics* is applied to problems dealing with particles of atomic or sub-atomic size. In essence the difference between these two approaches is that the classical one views matter and energy occurring as continuous functions whilst the quantum view is to visualize matter or energy as occurring in discrete quantities—*quanta*. According to the quantum theory atoms and indeed also molecules only exist in a discrete number of states, reflected in specific energy states; should there be a change of state then the atom or molecule must absorb or emit just the necessary energy to change it to another state. The concept of energy and energy states will be discussed in chapter 4; suffice it to demonstrate at this stage that certain changes in atomic structure can lead to the sudden and catastrophic liberation of energy—the atomic bomb. Atoms and molecules thus have (discrete) particular states with associated energy levels and these are described by sets of *quantum numbers*; before some aspects of the quantum theory of atomic structure can be described, the nature of electromagnetic radiation has to be outlined.

The basic measures which characterize waves can be illustrated by considering a long wire fixed to a wall and held taut (figure 2.2). Quick up and down movements of the hand holding the wire will generate a set of waves which will pass along the wire to the wall which may well reflect back the waves. The vertical distance between the top and bottom of the waves (length AB) is the *amplitude*, the distance between adjacent peaks (or troughs) (length CD) the *wavelength*, and the number of wave crests or troughs which pass a reference point in unit time is the *frequency*. The standard symbol for

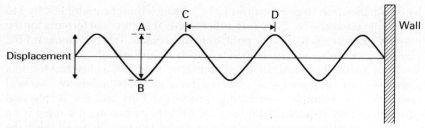

Fig. 2.2 A set of waves generated by the displacement of a string (exaggerated). Distance AB indicates amplitude, and CD, wavelength.

wavelength is the Greek letter lambda (λ) whilst for frequency, the symbol is the letter nu (v). These measures can be applied to all wave phenomena, from waves on the sea to the shortest known waves associated with cosmic rays. Such a range in values necessitates some mention of appropriate units. Wavelength (λ) is clearly a length measure and thus the conventions of the SI system must be followed. In the literature, measures of λ in ångströms (symbol Å) will be encountered, but these units are being discontinued ($1\text{Å} = 10^{-10}$ m). The unit in the SI system is the nanometre (1 nm = 10^{-9} m). In the electromagnetic spectrum, wavelength and frequency are related by the equation

$$c = \lambda v$$

where c is the velocity of light in a vacuum. Brief consideration of the units in this equation will indicate that frequency (v) has the dimension, per time unit (s^{-1}). The special name for this unit is the hertz (symbol Hz).

A sound wave results from the initial compression of the atmosphere whilst *electromagnetic radiation* is caused by the disturbance of electric and magnetic fields. The wavelengths and frequencies of the electromagnetic spectrum are illustrated in figure 2.3. Visible light constitutes only a very small part of this spectrum (ranging from 400–700 nm). A rainbow is an obvious example to quote in order to demonstrate that light is composed of a mixture of wavelengths, small ranges of wavelengths being associated with particular colours. The composition of white light is illustrated in figure 2.4.

An *emission spectrum* refers to the wavelength range associated with radiation generated by a source whilst an *absorption spectrum* is the range left after some wavelengths have been absorbed by passage through a medium. For example the spectrum of solar radiation which is received at the surface of the

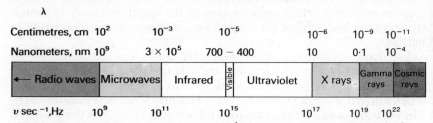

Fig. 2.3 The electromagnetic spectrum. The upper scales give the approximate wavelength ranges of the spectral regions in some common units. The lower scale gives the frequency in cycles per second or hertz. (after Gymer 1973, p. 33)

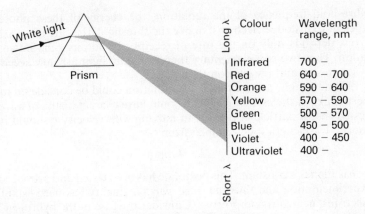

	Colour	Wavelength range, nm
Long λ ↑		
	Infrared	700 −
	Red	640 − 700
	Orange	590 − 640
	Yellow	570 − 590
	Green	500 − 570
	Blue	450 − 500
	Violet	400 − 450
	Ultraviolet	400 −
Short λ ↓		

Fig. 2.4 Dispersion of white light. (after Gymer 1973, p. 33)

earth is different to that which arrives at the top of the stratosphere; reflection from clouds and scattering of radiation by tiny particles in part explain this phenomenon, but radiation is also differentially absorbed in the stratosphere and troposphere. The same applies to the radiation emitted from the earth; much of the long-wave terrestrial radiation is absorbed by clouds, but certain wavelengths are able to pass through with little loss. Instruments known as *spectrometers* are used to determine emission and absorption spectra associated with samples of material. Such an approach has proved to be very important with regard to atomic and molecular structure.

The applicability of wave theory to radiation has been widely accepted, but it is also possible to consider radiation in terms of photon theory. Much of the work in this latter approach was pioneered by Planck in the early years of this century. He tried to explain, without success, the relationship between temperature and radiation (illustrated in figure 6.1 in chapter 6) using traditional classical physics; instead he proposed that matter which emitted or absorbed radiation could only occur in discrete energy states. He argued that the emission of such energy is resultant upon vibration within atoms or molecules and proposed the following relationship between energy (E) and frequency (v):

$$E = nhv$$

where n is whole numbers, 0, 1, 2, 3, ... and h is a constant—later called Planck's constant. Since n is always a whole number, then discrete energy values are obtained. This is the foundation of quantum theory; the emission (and absorption) of radiation by atoms or molecules is only possible at definite energy levels—transitional states are impossible (Gymer 1973). Einstein developed these ideas and he proposed that all radiation is composed of distinct energy packets or quanta; this is the basis of photon theory which can be demonstrated by the photoelectric effect. Light shining on certain metals in particular situations causes the loss of electrons. Such emission of electrons only takes place once a threshold frequency of radiation is achieved, irrespective of intensity. Above this frequency, the rate of release of electrons is only dependent upon the intensity of the radiation. This radiation can be thought of as a stream of photons, each carrying a quantity of energy. Below the threshold frequency of the radiation, the energy of these photons is insufficient to dislodge electrons; above the threshold, the rate of emission of electrons depends only on the rate of receipt of photons—the intensity of radiation. Thus two complementary theories about light are now accepted—the wave theory and the photon theory.

In 1924 de Broglie argued that since radiation could be considered to have properties like particles, then particles could have characteristics of waves. He suggested that a particle of mass m in moving with velocity v should have a diagnostic wavelength λ determined from

$$\lambda = h/mv$$

where h is Planck's constant. This postulate has been tested and proved correct by experimentation and thus the *wave–particle* duality has been established for electrons, neutrons and protons. Consider the case of the hydrogen atom with one electron encircling the nucleus. It can now be appreciated that this electron does not move in a simple circle, but instead has a complex wave path. One important point about the wave path of the electron round the nucleus is that there will be one particular wavelength for one mean distance from the nucleus if the wave-like movement of the electron is to be in phase. Consider figure 2.5: there will be one wavelength such that the wave pattern can join up in continuous fashion round the nucleus. In other words there must be an integral number (n) of wavelengths (λ) which fit this circle (radius r). Thus

$$2\pi r = n\lambda = nh/mv$$

The integral number n is an example of a quantum number since it in part describes the movement of the electron.

The conception of atomic structure by Bohr was possible because he realized that atoms, when subjected to an input of energy, respond in producing a distinctive pattern of radiation. In fact each element has its own spectrum which is *not continuous* as with the electromagnetic spectrum, but

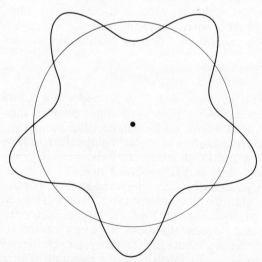

Fig. 2.5 The wave pattern of an electron round a nucleus (exaggerated).

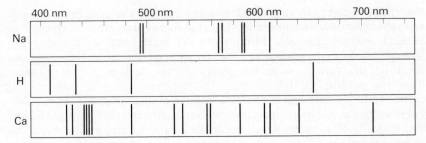

Fig. 2.6 Spectra for sodium, hydrogen and calcium. (after Gymer 1973, opposite p. 40)

instead is discrete since emission is only at a few specific wavelengths. For example if an enclosed cylinder is full of hydrogen and the gas is subjected to a high voltage, then a characteristic spectrum of a series of lines corresponding to particular wavelengths can be produced by spectographic analysis (figure 2.6); four of these lines occur within the wavelength range 400–700 nm, a violet, a blue violet, a blue green and a red-orange line. Figure 2.6 also shows spectra for sodium and calcium. In effect each atomic structure has a characteristic spectrum akin to a finger print and such a principle is of tremendous practical importance since the elemental analysis of soils, rocks or gases by spectographic techniques is thus possible. Bohr's view of the atom was that electrons encircled the nucleus in orbit-like fashion; an input of energy led to its subsequent emission by the exchanging of orbits by electrons, for example the red-orange line in the case of hydrogen results from electrons changing from the third orbit to the second.

As already described, Bohr's model of electron movement round nuclei similar to planetary movement was revolutionized by the ideas of de Broglie who suggested that electrons move in a wave path in a resonant manner. Several other fundamental contributions were made during the inter-war years to the understanding of atomic structure, but besides the work of de Broglie, only that of Heisenberg will be mentioned. He introduced an *uncertainty principle* which has implications not only for electronic structure, but also for such themes as order and disorder in environmental systems and how these systems evolve through time. Heisenberg reasoned that any measurement of a variable within a system necessarily causes a disturbance; the more precise the measurement the greater will be the associated disturbance. This means that measurement at the atomic scale can never completely describe the nature of the atom, or in other words there is a degree of uncertainty about the system. For example, the precise behaviour of an electron cannot be specified, instead its behaviour can only be described in terms of probability. The behaviour of the electron can be viewed as occurring most likely within a tolerance zone similar to confidence limits and the degree of tolerance is specified in the quantum number.

The implication of Heisenberg's uncertainty principle is that the exact trajectories of electrons cannot be determined. It is thus wrong to describe the movement of electrons as following fixed orbits (the Bohr model), but instead to think in terms of a more general orbital spatial pattern to accommodate the degree of uncertainty. As already described, electrons move in orbital paths within prescribed shells round the nucleus. It has also been shown that the

B

absorption of energy by atoms leads to a movement in electrons between shells and subsequent emission of energy. Thus the energy of an atom is expressed in its electronic configuration. Electrons in the nearest shell to a nucleus are strongly bound to the nucleus and thus are difficult to remove. Hence such an electronic structure possesses low energy. As the shells at greater distances from the nucleus fill up, so they become more easy to remove and correspondingly have higher energies. A large number of shells are possible, but possible electronic configurations within the first five shells account for the vast majority of elements which occur within the natural environment. An important principle is that shells are filled with electrons from the inner side in sequence from the innermost shell and also that the individual shells have finite limits in the numbers of electrons which they can accommodate. These are as follows:

$$\begin{array}{lll}
\text{1st shell} & : & 2 \text{ electrons} \\
\text{2nd shell} & : & 8 \text{ electrons} \\
\text{3rd shell} & : & 8 \text{ electrons} \\
\text{4th shell} & : & 18 \text{ electrons} \\
\text{5th shell} & : & 18 \text{ electrons} \\
\text{6th shell} & : & 32 \text{ electrons}
\end{array}$$

Consider the hydrogen atom ($_1^2$H) which has one electron which must go into the first shell and thus to leave a vacant space in this shell. Helium ($_2^4$He) has two electrons and thus fills the first shell. The next element is lithium ($_3^6$Li) with three electrons; the first shell is filled with two electrons but the third electron is forced to occur in the second shell. Oxygen ($_8^{16}$O) has two electrons in the first shell and six in the second to leave two vacancies in the outer shell. The electronic structure of the first 25 elements is presented in table 2.2.

The number of electrons or the number of vacancies in the outer shell is of prime importance since the number of bonds between atoms is dependent on such numbers. The number is known as the *valency* of an element; it is usual for the valency to be the smaller number either of electrons in an outer shell or of electron vacancies in the shell. The number of vacancies in a shell is obtained by subtracting the actual number of electrons from the maximum possible number. From table 2.2 it can be seen that hydrogen, lithium, fluorine, sodium, chlorine and potassium are of valency one. Examples of elements with valency two are beryllium, oxygen, magnesium, sulphur and calcium although some elements (for example, sulphur) can display more than one valency, a result of their ability to have a varying number of electrons in their outermost shells. The electronic configuration of certain elements in table 2.2 means that their shells are complete (helium, neon, argon); these are examples of the inert gases. Chemical reactions involve the exchange of electrons and thus the valency of an atom can be thought of as a measure of its combining capacity.

Electronic configuration for individual elements appears a complicated topic, and indeed there are many other aspects of the subject which have not even been hinted at in this chapter. However, order in electronic configuration between various elements is apparent when the nature of the *periodic table* is outlined. The fundamental contribution of Mendeleev and Meyer was to set out elements horizontally in order according to atomic number and at the same time to arrange them vertically so that similar elements were arranged adjacent to each other (appendix 3). The importance of the number of

Table 2.2 Distribution of electrons in shells for the first 25 elements.

Element	Symbol	Atomic number (Z)	Number of electrons			
			1st shell	2nd shell	3rd shell	4th shell
Hydrogen	H	1	1			
Helium	He	2	2			
Lithium	Li	3	2	1		
Beryllium	Be	4	2	2		
Boron	B	5	2	3		
Carbon	C	6	2	4		
Nitrogen	N	7	2	5		
Oxygen	O	8	2	6		
Fluorine	F	9	2	7		
Neon	Ne	10	2	8		
Sodium	Na	11	2	8	1	
Magnesium	Mg	12	2	8	2	
Aluminium	Al	13	2	8	3	
Silicon	Si	14	2	8	4	
Phosphorus	P	15	2	8	5	
Sulphur	S	16	2	8	6	
Chlorine	Cl	17	2	8	7	
Argon	A	18	2	8	8	
Potassium	K	19	2	8	8	1
Calcium	Ca	20	2	8	8	2
Scandium	Sc	21	2	8	8	3
Titanium	Ti	22	2	8	8	4
Vanadium	V	23	2	8	8	5
Chromium	Cr	24	2	8	8	6
Manganese	Mn	25	2	8	8	7

electrons in the outermost shell in conditioning the nature of elements has been stressed. Thus elements which have complete shells (the inert gases) are chemically very similar, as are elements having one electron outside a full shell (sodium and potassium). Such similarities are expressed in the structure of the periodic table. The rows of elements in the table are called *periods* according to atomic number: each new row begins after an element with full shells. Vertical columns of elements in the periodic table are known as *groups*; these elements have similar chemical properties because their electronic configurations are akin. Group I elements are distinguished by having one electron outside a closed shell, whilst elements in group VI (for example) need two electrons to complete a shell. The inert gases constitute group 0, groups I and II contain elements with electrons extending only to the third shell. Group III has far more elements because the fourth shell can contain up to 18 electrons. The elements in the middle of this group have numbers of electrons in this shell far removed from either the full or empty situation; thus they are called *transition* elements. It should be noted that hydrogen is not assigned to any group because of its unique electronic structure. In general terms the periodic table is extremely useful for ordering a wide variety of chemical and physical properties of elements.

A return can now be made to the question of chemical formulae; as already stated this can be obtained by chemical analysis, but it is far more common to postulate a formula on the basis of the periodic table. For example, if it is known that one cation of sodium combines with one anion of chloride to form sodium chloride with the formula NaCl, then it is very likely that the formula for potassium chloride will be KCl given the fact that Na and K are adjacent in the same column of the periodic table. A similar argument could be put forward for sodium oxide and potassium oxide, Na_2O and K_2O; in this case two cations of Na and K combine with one anion of oxygen. This is understood by recalling that Na and K have an isolated electron in their outermost shell— hence one single positively charged electron is available for bonding per atom; in contrast the O atom has two vacancies in its second shell and thus two additional electrons must be added in order for the shell to be electronically complete. Such a deficiency in electrons means that the O atom is an anion and requires to be bonded with two cations if these have only one available electron. Examples of monatomic cations are Na^+ (sodium), Ca^{2+} (calcium) and Al^{3+} (aluminium); the number of + signs indicates the number of excess electrons in the outer shell available for bonding. In certain cases it is possible for a metal cation to exist in more than one electronic state; for example Fe^{2+} (iron II) and Fe^{3+} (iron III), Cu^+ (copper I) and Cu^{2+} (copper II), and Sn^{2+} (tin II) and Sn^{4+} (tin IV). The old convention is to describe Fe^{2+} as ferrous and Fe^{3+} as ferric—similarly cuprous and cupric, stannous and stannic. The number of charges, be they positive or negative, corresponds to the valency of the atom. In the physical environment it is very common for ions to be made up of several atoms—polyatomic ions. Ammonium (NH_4^+) is a cation with one positive charge resultant upon the linkage of four atoms of hydrogen with one of nitrogen. Examples of polyatomic anions are:

ClO_3^-	chlorate
OH^-	hydroxide
NO_3^-	nitrate
NO_2^-	nitrite
CO_3^{2-}	carbonate
SO_4^{2-}	sulphate
SO_3^{2-}	sulphite
CrO_4^{2-}	chromate
$Cr_2O_7^{2-}$	dichromate
PO_4^{3-}	phosphate

A few examples will illustrate the derivation of formulae from combinations of these ions. A useful first step is just to write the symbols of the ions with the associated charges. Consider calcium chloride, $Ca^{2+}Cl^-$; such an expression is not electrically balanced and thus 2 anions of Cl^- must be bonded to one of Ca^{2+} in order to obtain such a balance and thus the formula is $CaCl_2$ with the subscript 2 indicating 2 anions of chlorine linked to 1 (assumed) of calcium.

Other examples

$$\text{iron (III) oxide } Fe^{3+}O^{2-} \quad Fe_2O_3$$

(A useful tip with this example is to 'cross-multiply' the charges in order to obtain the subscripts.)

$$\text{ammonium carbonate } NH_4^-CO_3^{2-} \quad (NH_4)_2CO_3$$

(Note that NH_4 must be enclosed in brackets.)

iron (II) sulphate Fe^{2+} SO_4^{2-} $FeSO_4$

(Note in this instance that both ions are of valency 2 and thus no balancing is required.)

It can thus be seen that a chemical formula is a quantitative method of indicating the composition of a molecule. The approach is also applicable to lattice structures such as sodium chloride since the ions are held together by forces of ionic bonding. It is often useful when dealing with more complicated structures to represent them in two dimensions. For example $(NH_4)_2SO_4$ (ammonium sulphate) can be represented as follows:

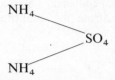

In this instance the individual bonds are represented by the rules. The practical significance of such an approach will be exemplified with respect to water in chapter 7 since certain fundamental properties of water are explicable in terms of its molecular structure:

$$\underset{H}{\overset{O}{\diagdown}}\,\underset{H}{\diagup}$$

In organic chemistry the need to represent compounds in this manner becomes very marked; in fact the use of three-dimensional hardware models is most suitable. Carbohydrates are compounds of carbon, hydrogen and oxygen, and a simple example is glucose with the molecular formula $C_6H_{12}O_6$. More information about the molecular structure of glucose is available if it is represented in two dimensions, for example:

$$O = \underset{|}{C} - \underset{\underset{OH}{|}}{\overset{\overset{H}{|}}{C}} - \underset{\underset{OH}{|}}{\overset{\overset{H}{|}}{C}} - \underset{\underset{OH}{|}}{\overset{\overset{H}{|}}{C}} - \underset{\underset{OH}{|}}{\overset{\overset{H}{|}}{C}} - \underset{\underset{OH}{|}}{\overset{\overset{H}{|}}{C}} - \overset{\overset{H}{|}}{C} - H$$

A quick check of this diagram will show that the different valencies of H, C, O and OH have been taken into account. This is an example of a chain structure; several other types of structures are possible ranging from the simplest structures such as hexagons (e.g. benzene) to double helixes (e.g. DNA). These are best demonstrated in three-dimensional models.

Of prime concern to physical geographers are the processes of chemical reactions—weathering processes are obvious examples. A chemical reaction necessarily involves a change in chemical composition and the whole process can be neatly and quantitatively expressed by a chemical equation. This expresses the course of the reaction from the initial constituents to the final product(s). If $CaCO_3$ (calcium carbonate) is added to water in which CO_2 (carbon dioxide) has been dissolved, then $Ca(HCO_3)_2$ (calcium bicarbonate) will form. This process can be represented as follows:

$$CaCO_3(s) + CO_2\ (aq) + H_2O \rightarrow Ca(HCO_3)_2\ (aq)$$

The arrow indicates the direction to the process and in this instance the arrow could easily point in the opposite direction since this process is an example of an easily reversible reaction. The letters in parentheses indicate the states of the substances: s, solid; aq, dissolved in water; other possibilities are l, liquid and g, gaseous. This information is not always given in equations. Since matter can neither be created nor destroyed, the products on the right-hand side of the equation must equal the ingredients to the reactions; in other words the same number of atoms of the different elements must occur on either side of the equation. The above equation is balanced since this condition is satisfied, otherwise the equation would have to be modified by trial and error until a balanced equation could be produced. This can be demonstrated by an example which will also illustrate the quantitative nature of a balanced equation.

Example

Coal can contain the impurity iron pyrite (FeS_2) which during the combustion process combines with oxygen to produce iron (III) oxide and sulphur dioxide. The first step is to write out the equation in an unbalanced form:

$$FeS_2 + O_2 \rightarrow Fe_2O_3 + SO_2$$

As an initial attempt the number of Fe atoms can be balanced by inserting a 2 in front of FeS_2; this means that there must be 4 molecules of SO_2 to balance S atoms:

$$2FeS_2 + O_2 \rightarrow Fe_2O_3 + 4SO_2$$

This equation is balanced thus for Fe and S atoms, but there are 11 O atoms on the right-hand side and thus the O_2 on the left-hand side would need to be multiplied by $5\frac{1}{2}$ to balance this equation. Molecules occur in whole numbers and thus this is impossible. The final solution is therefore obtained by multiplying by another factor of 2:

$$4FeS_2 + 11O_2 \rightarrow 2Fe_2O_3 + 8SO_2$$

This equation can now be used to predict how much SO_2 will be produced from the combustion of a given quantity of iron pyrite. In essence the equation states that 4 molecules of FeS_2 combine with 11 molecules of O_2 to produce 2 molecules of Fe_2O_3 and 8 of SO_2. Of course concern could be with any constant multiple of these numbers of molecules and a convenient one to select is Avogadro's number ($6 \cdot 022 \times 10^{23}$) which is the number of molecules in one mole. Thus the numbers in front of the terms in the equation can be taken to correspond to moles. This means that the combustion of 4 mol of FeS_2 results in 8 mol of SO_2 or 1 mol of FeS_2 results in 2 mol of SO_2. The next step is to determine the mass of a mole of FeS_2 and SO_2—the gram molecular weights. Now 1 mol of FeS_2 can be determined by adding 1 gram atomic weight of Fe ($55 \cdot 84$ g) and 2 gram atomic weights of S ($32 \cdot 06$ g), which is ($55 \cdot 84 + 2 \times 32 \cdot 06$) g, to equal $119 \cdot 96$ g. The same type of calculation for 1 mol of SO_2 is obtained by adding 1 gram atomic weight of S ($32 \cdot 06$ g) and 2 gram atomic weight of O ($16 \cdot 00$ g) to give $64 \cdot 06$ g. Thus the combustion of $119 \cdot 96$ g of FeS_2 produces ($64 \cdot 06 \times 2$) g of SO_2 and hence $1 \cdot 00$ g of FeS_2 releases $1 \cdot 07$ g of SO_2. Suppose a power station is burning coal which contains iron pyrite and every hour 100 g of iron pyrite are consumed. Then, if no precautions are taken, 107 g of SO_2 will be released into the atmosphere every hour. Similar quantitative analyses are possible for any chemical reaction for which a balanced equation is available.

For an understanding of weathering processes it is not only necessary to be aware of the nature of reactions, but also to know why there is a reaction and to appreciate the factors influencing the rates of the reactions. Such a desire demands a knowledge of reaction energetics since a chemical results from a change in free energy. Rather than describing the various types of chemical reactions at this stage, it seems appropriate to postpone further discussion of this topic until the principles of energy are described. In this chapter emphasis has been given to the atomic nature of matter and it may be felt that such an approach is only of relevance to chemistry. Such a view would be very wrong since the physical nature and properties of matter can only be understood with reference to the basic building blocks of matter in the physical environment—atoms.

3

Foundations of physics

The previous chapter indicated that a knowledge of quantum mechanics is fundamental to an understanding of the atomic nature of matter. As was explained, the development of quantum mechanics came through the realization that the principles of classical mechanics could not be applied at the scale of individual atoms. In contrast, problems at the macroscopic level demand the use of classical or Newtonian physics. For example when the processes of a landslip, or atmospheric circulation or soil creep are being studied, then the principles of classical physics are very relevant. This chapter describes and illustrates the nature of the SI derived units in physics mentioned in chapter 1 since it is important for the physical geographer to have a firm understanding of such units as velocity, acceleration, force, pressure, work, power and energy. Concepts associated with heat and with the particular physical states are left to chapters 6 and 7 respectively.

Before systematic attention is given to these measures, it is useful to note that some of them are scalar quantities whilst others are vectors. The only attribute of a scalar quantity is its magnitude; in contrast a vector also possesses direction. The base unit, mass, is clearly a scalar quantity whilst a force is a vector given its magnitude and direction attributes. Common scalar and vector quantities encountered in physics are listed below:

Scalar	Vector
distance	displacement
speed	velocity
mass	acceleration
work	force
energy	momentum

The graphical representation and resolution of vectors will be demonstrated in this chapter with reference to forces, but the reader should also be acquainted in greater detail with the nature of vector algebra (see Sumner 1978, chapter 2).

3.1 Length, velocity and acceleration

The dimension length always has magnitude and can have direction. For instance the length of a specific contour in a drainage basin can be stated as 1 km, no direction being implied. The distance between two grid references has magnitude as well as direction since the length measure is tied into a frame of reference. Suppose a particle is at position P_1 on an X-axis at time t_1 and at P_2, also on the X-axis at time r_2 (figure 3.1). The distance between P_1 and P_2 can be

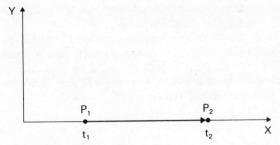

Fig. 3.1 Particle moves along X-axis from P_1 to P_2.

represented as ΔX and the corresponding time interval is Δt obtained by subtracting t_1 from t_2. The average velocity of the particle can then be determined by dividing the distance (ΔX) by the time (Δt), i.e.

$$\bar{v} = \frac{\Delta X}{\Delta t} \text{ where } \bar{v} \text{ represents average velocity.}$$

The mean velocity clearly is a vector quantity since it has magnitude and direction even though the latter does not change. A curved path would necessitate such a change (figure 3.2). In this case the particle follows a curved path from P_1 to P_2 over the interval of time Δt; let the distance along the path between P_1 and P_2 be s then the average velocity ($\bar{v}$) is equal to $s/\Delta t$. The direction of movement between P_1 and P_2 will change as the particle moves. At any intervening position, say P the direction of movement will be tangential to that point. The average velocity of the particle between P_1 and P_2 may well conceal variations in velocity over that distance. To obtain the velocity at specific point P, the instantaneous velocity, the distance travelled in a very short period (Δs) must be divided by the time interval (Δt), i.e.

$$v = \lim_{\Delta t \to 0} \frac{\Delta s}{\Delta t} = \frac{ds}{dt}$$

This states in mathematical form that velocity is the rate of change of distance with time. Again, instantaneous velocity has magnitude and direction, thus it is

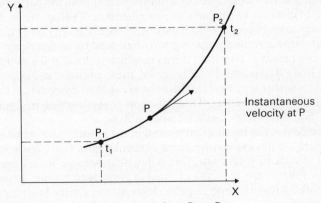

Fig. 3.2 Particle moves along a curved path from P_1 to P_2.

a vector. The distinction between velocity and speed should now be evident since the latter only implies magnitude.

If the velocity of the particle is varying between P_1 and P_2, it may be of interest to determine the rate of change of velocity—the property known as acceleration. Exactly the same argument could be presented as above for acceleration in order to distinguish between average acceleration ($\bar{a}$) and instantaneous acceleration (a). Hence

$$a = \lim_{\Delta t \to 0} \frac{\Delta v}{\Delta t} = \frac{dv}{dt}$$

where Δv represents the change in velocity.

3.2 Forces

A force is defined in terms of moving a mass with a particular acceleration; in the SI system the unit of force is the newton (N) which is the force necessary to give a mass of 1 kg an acceleration of 1 m s^{-2}. Dynes will also be encountered and 1 dyne equals 10^{-5} N. A force (F) can be determined by multiplying the mass of a body (m) by its acceleration (a), i.e.

$$F = ma$$

This can be rewritten as:

$$\frac{F}{a} = m$$

to indicate that mass can be defined in terms of force per unit of acceleration. Acceleration is a vector whilst mass is a scalar quantity and therefore a force is also a vector.

As already indicated, weight is not a measure of mass according to the SI system; the weight of a body is the *force* exerted on it by the gravitational effect of the earth. Thus a mass m is attracted to the earth by the force mg where g is the acceleration due to gravity. When a golf ball is dropped from a height it accelerates towards the ground at a constant rate of 9·81 m s^{-2} assuming air resistance is negligible. Thus the force attracting the ball to the ground is its mass times this acceleration. It should be stressed that gravity is an acceleration and its value varies to a limited extent over the earth because of differences in distance to the earth's centre, because of variations in density of crustal material and because of the rotation of the earth. Any body resting on a surface has its force resultant upon gravity balanced by an equal and opposite force, a consequence of Newton's third law which states that if a body A exerts a force on body B, then this latter body will exert an equal and opposite force on body A. Another way to state this is to say that to every action there is an equal and opposite reaction. This situation is represented in figure 3.3 to illustrate these two forces resultant upon a mass m.

All movement is the result of forces and in the natural environment at least several forces are always involved in a particular situation. There is need to simplify a system of forces and this is possible because forces are vectors. Consider a simple case of a ball of mass m suspended by one string from a ceiling (figure 3.4 (a)); the string exerts a force mg to keep the ball suspended. It would also be possible to achieve exactly the same effect by suspending the ball

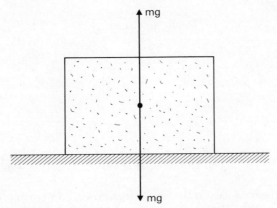

Fig. 3.3 The force resultant upon the mass *m* of a body resting on a surface is exactly countered by the force which the surface exerts on the body.

by two strings as in figure 3.4 (b). In other words the two forces F_1 and F_2 have exactly the same net effect as the force on the single string in figure 3.4 (a). Suppose forces F_1 and F_2 are at angles θ_1 and θ_2 to the vertical. Any vector can be resolved into component vectors; in this example consideration is limited to two dimensions and thus the forces F_1 and F_2 can each be resolved into two components. Imagine that figure 3.4 (b) is drawn to scale, in other words the lengths of the arrows are proportional to the magnitude of the forces and true directions are also indicated. Then in order to determine the magnitude of F_1 on the vertical line, the length of the projection of F_1 on this line must be determined. In other words it is a simple trigonometrical problem to calculate the vertical component of F_1—say this is called F_{1y}. Then

$$\cos \theta_1 = \frac{F_{1y}}{F_1}$$

and hence
$$F_{1y} = F_1 \cos \theta_1$$

Similarly
$$F_{2y} = F_2 \cos \theta_2$$

The horizontal resolutions of these forces are $F_{1x} = F_1 \sin \theta_1$ and $F_{2x} = F_2 \sin \theta_2$ and these are equal and opposite in direction. This is a consequence of the ball being in equilibrium since in this situation the sum of the horizontal components of the forces must equal zero; the same applies of course to the vertical components. For this latter situation in the example, the downward

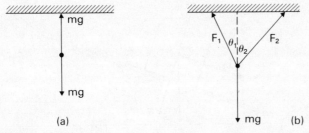

(a)

(b)

Fig. 3.4 The forces associated with a sphere of mass *m* on being hung by one string in (a) and by two strings in (b).

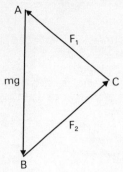

Fig. 3.5 The triangle of forces which are shown in Figure 3.4(b).

force *mg* must be balanced by the sum of the vertical components of F_1 and F_2. Hence

$$mg = F_{1y} \cos \theta_1 + F_{2y} \cos \theta_2$$

If the lengths of the strings from the ceiling are equal, F_1 will equal F_2 and θ_1 will be the same as θ_2. In this situation:

$$mg = 2F\cos\theta$$

The same results could have been obtained by using vector algebra but it is useful to note that vectors can also be analysed by graphical procedures. The three forces as shown in figure 3.4 (b) can be indicated by a triangle of forces as in figure 3.5: side AB is proportional in length to force *mg* and indicates the direction of the force; the same applies to BC for force F_2 and CA for F_1. This is the graphical addition of vectors and can be applied to more complex situations to produce polygons of vectors.

If a block of mass *m* is placed on a plank in a horizontal position, then no movement will take place because the sums of the force components in both the horizontal and vertical directions are zero. However, as the plank is tilted up at one end, at first the block will stay in its same position, but there will be found a critical angle of inclination of the plank when the block begins to move. For the analysis of such a problem, a discussion of frictional forces must be introduced. When one body slides over another, each exerts a frictional drag which is parallel to the surfaces. Each restraining force will be opposite to the direction of movement. Figure 3.6 illustrates the situation of the plank and the block of mass *m* when the block just begins to move; the plank is inclined at angle θ. The

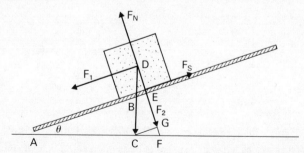

Fig. 3.6 The forces associated with a block resting on an inclined surface.

forces can be resolved into two directions, parallel to the surface of the plank and at right angles to it; thus the force resultant upon the mass of the block can be resolved into F_1 and F_2, the magnitudes of which can be determined from triangle DCG. Triangle AEF has one right-angle and thus angle AFE has a value of $(90° - \theta)$. Similarly triangle CDF has one right-angle and thus angle CDF plus angle AFE must equal 90°. But angle AFE has a value of $(90° - \theta)$ to mean that angle CDF has a value of θ:

$$\sin \widehat{CDG} = \sin \theta = \frac{F_1}{mg}$$

(g is acceleration due to gravity)

$$\therefore F_1 = mg \sin \theta$$

Similarly

$$F_2 = mg \cos \theta$$

The equal and opposite force to F_2 is the force which the plank exerts to maintain the block on its surface (F_N). Along the surface of the plank, the downslope force component F_1 must be equal to the frictional force, F_s, at the instant when slippage is about to occur. At this instant:

$$F_s = F_1$$

But

$$F_1 = mg \sin \theta$$

The value of F_s will vary according to the force normal to the surface and according to the nature of the surface between the block and plank. There is need to derive some measure of friction which is exclusive of the normal force and this is achieved by defining the coefficient of friction as the ratio of the force required to overcome friction to the normal force. Thus

$$\mu_s = \frac{F_s}{F_N} = \frac{F_1}{F_2} = \frac{mg \sin \theta}{mg \cos \theta} = \tan \theta$$

where μ_s is called the *coefficient of static friction*. The critical angle when movement just takes place is called the angle of repose and is represented as θ_s. After the block has started to move, it might be possible to lower the plank to another particular angle which would allow the block to continue to move downslope at constant velocity. In other words once movement has been initiated, a smaller downslope force component is necessary in order to maintain movement at constant velocity. A second coefficient of friction is thus required—μ_k with the subscript k to indicate the dynamic state. Exactly the same argument could be presented as above to show that $\mu_k = \tan \theta_k$. The distinction between μ_s and μ_k should be evident given the difference between the pull necessary to start a sledge moving and the pull needed to keep it on the move. It is the coefficient of static friction which tends to be of greater importance to such problems as blocks of rock sliding down hillslopes. The critical aspect is the initiation of movement—the same applies to landslip problems. The dynamic aspects of friction are of relevance to phenomena which are always on the move—for example water in rivers or gases in the atmosphere. The frictional measure of gases or fluids is known as *viscosity*, but a description of this must be postponed until chapter 7 since the principles of shear stresses have yet to be outlined.

Other forces of importance in the physical environment result from circular motion; an understanding of these is necessary in order to build explanatory models of atmospheric circulation. Consider a particle revolving in a circle of radius r (figure 3.7), and let points P_1 and P_2 indicate two positions of the particle. Over an interval of time (Δt) the particle moves from P_1 to P_2 and the vector describing such overall movement is represented by Δs. Thus the average velocity ($\bar{v}$) is obtained by dividing Δs by Δt. The instantaneous velocity at any intervening point can be determined by obtaining the limit of $\Delta s/\Delta t$ as Δt tends to zero. Thus at the intervening point P_3 the velocity is given by ds/dt and the direction of v is tangential to the circle at that point. Suppose that the particle has a velocity of constant magnitude and that one complete revolution ($2\pi r$) is completed in time T. Thus the magnitude of the constant velocity is $2\pi r/T$ though the direction of the velocity is always changing.

If the magnitude of velocity round the circle is not constant, then the particle must be accelerating or decelerating. Let v_1, a vector, be the velocity at P_1 and v_2 the velocity at P_2; thus a (acceleration) must be the vector difference ($v_2 - v_1$) divided by the interval of time Δt, i.e.

$$a = \frac{v_2 - v_1}{\Delta t}$$

or

$$a\Delta t = v_2 - v_1$$

In order to determine the direction of acceleration, a triangle of vectors can be constructed as in figure 3.8. The two vectors v_1 and v_2 are drawn from point Q ensuring that the lengths and direction of these vectors are the same as P_1 and P_2 respectively. The need is to determine ($v_2 - v_1$) and thus $-v_1$ is indicated by reversing the direction of the arrow. The addition of vectors v_2 and $-v_1$ is indicated by side RS of the triangle: this can be explained by stating that a trip from R to S via Q has the same effect as going direct from R to S. But ($v_2 - v_1$) has been determined as ($a\Delta t$) and hence the direction of the acceleration vector is roughly towards the centre of the circle. With constant speed, the magnitude of v_1 will equal the magnitude of v_2; the particle in moving from P_1 to P_2 sweeps an angle of θ at the centre of the circle and this must be the same as the difference in orientation of the vectors v_1 and v_2. As the angle θ tends to zero, so the direction of vector ($a\Delta t$) will be increasingly towards the centre of the circle.

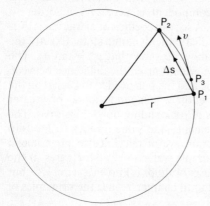

Fig. 3.7 A particle, in following a circular path, moves from P_1 to P_2.

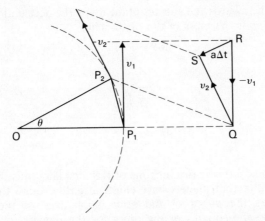

Fig. 3.8 The construction necessary to determine the acceleration of the particle in Figure 3.7.

The magnitude of the acceleration vector is also of importance. It has already been established that triangles OP_1P_2 and QRS must be similar since they both have two sides of equal length with the same angle in between. The distance of P_1P_2 must be $(v\Delta t)$ where v is the value of the constant speed. Thus by proportions,

$$\frac{v\Delta t}{r} = \frac{a\Delta t}{v}$$

and hence

$$\frac{v}{r} = \frac{a}{v}$$

or

$$a = \frac{v^2}{r}$$

The important conclusion is that the acceleration of a body which is moving in a circle is towards the centre of the circle and has the magnitude v^2/r. Newton's second law states that the acceleration of a body is directly proportional to the resultant force acting upon it and is inversely proportional to the mass of the body. The existence of an acceleration towards the centre of a circle with circular motion then implies there is a net force in that direction. Since $F = ma$, and $a = v^2/r$, then $F = mv^2/r$. This acceleration and force, since they are directed to the centre, are known as *centripetal*. Newton's third law states that to every action there is an equal and opposite reaction—the force away from the centre of the circle is known as the *centrifugal* force. If a stone is tied to a string and rotated, the string exerts a pull on the stone towards the centre of the circle of rotation (the centripetal force), whilst the stone exerts on the string an equal and opposite force (the centrifugal force).

These principles can be applied to planetary movement. Suppose that the earth moves round the sun in a circular orbit, then the earth must have an acceleration towards the sun of (v^2/r) where v is the speed of the earth's movement and r is the distance from the sun to the earth. The length of the earth's orbit is $(2\pi r)$ and hence the speed of the earth is $(2\pi r/t)$ where t is the time of one complete revolution of the sun by the earth. The centripetal force

(F) of the earth towards the sun resultant upon the acceleration of (v^2/r) is (mv^2/r), where m is the mass of the earth. Thus

$$F = \frac{mv^2}{r}$$

$$= m\left(\frac{2\pi r}{t}\right)^2 \frac{1}{r}$$

$$= \frac{4\pi^2 mr}{t^2}$$

In order to develop this reasoning, one of Kepler's laws must be introduced. Kepler recognized that planets have elliptical paths round the sun and he discovered that the cubes of the semi-major axes of the ellipses are proportional to the squares of the times for the planets to complete one revolution. This means that t^2 is proportional to r^3, which can be written as $t^2 = kr^3$ where k is a constant.

The above equation can thus be rewritten as

$$F = \frac{4\pi m}{k} \frac{1}{r^2}$$

which can be simplified to

$$F = c\frac{m}{r^2} \text{ where } c \text{ is a new constant.}$$

The force is thus proportional to the mass of the earth and inversely proportional to the square of the radius of its orbit. If it can be argued that the mass of the earth has an effect on the force between the earth and the sun, it can be similarly argued that the mass of the sun must also have an influence. This is the foundation to Newton's law of gravity which states that between every two particles there is a force of gravitational attraction which is proportional to the product of their masses and inversely proportional to the square of the distance between them, i.e.

$$F_G = G\frac{m_1 m_2}{r^2} \tag{3.1}$$

where F_G is the gravitational force, m_1 and m_2 are the masses, r the distance between them and G the universal gravitational constant which is equal to $6 \cdot 67 \times 10^{-11}$ N m^2 kg^{-2}. The force of attraction between masses on the earth's surface is infinitesimally small and is rarely of relevance; at the atomic level these principles of classical mechanics are inapplicable. Equation (3.1) has, however, been taken as the basis for gravity models in human geography whereby the attraction between two settlements of populations P_1 and P_2 is inversely proportional to the square of their distance apart. Clearly Newton's law of gravitation is of profound importance to an understanding and thus prediction of planetary and satellite movement. One field expression of this law was found by the early surveyors in northern India when they found that their plumb-bobs were slightly deflected away from the vertical towards the Himalayas.

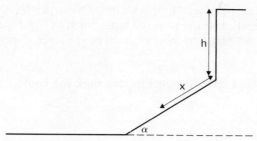

Fig. 3.9 The parameters to describe the height of a free face, the length which a fallen boulder slides down a scree slope and the angle of the scree slope.

A worked example seems appropriate at this stage to illustrate some of these physical principles. The selected example is that of a scree slope rockfall model and Statham's (1976) analysis is amplified for this purpose. Imagine a free rock face which has a scree slope inclined at its base (figure 3.9). Suppose that a boulder is loosened at the top of the free face and falls a distance h to the top of the scree slope which is inclined at an angle α to the horizontal. The boulder then slides down a distance x before coming to rest. The boulder falls with an acceleration g and its velocity (v_f) when it hits the top of the scree slope is obtained from the following equation:

$$v_f^2 = u_f^2 + 2gh \tag{3.2}$$

where u_f is the initial velocity, equal to zero. Thus

$$v_f^2 = 2gh \tag{3.3}$$

The boulder hits the top of the scree slope with velocity v_f, and begins to slide downslope, initially with the downslope component of velocity of v_f; let this downslope initial velocity be u_s, which can be obtained by considering figure 3.10. Line AC represents vector v_f whilst lines BC and AB represent the velocity components parallel to and at right-angles to the scree slope (AD).

$$\sin \widehat{BAC} = \sin \alpha = u_s/v_f$$

$$\therefore u_s = v_f \sin \alpha \tag{3.4}$$

Intuitively this seems correct since if there was no scree slope ($\alpha = 0$), there would be no downslope velocity component. The boulder then starts sliding on the scree slope, but is brought to rest at distance x because of friction. A

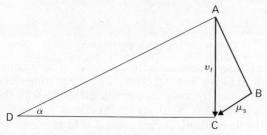

Fig. 3.10 Resolution of velocity vectors for the rockfall example.

deceleration thus operates to retard its movement; let this downslope deceleration (negative acceleration) be represented by $-f$. Equation (3.2) can be re-written appropriate to movement down the scree slope:

$$v_s^2 = u_s^2 - 2fx$$

where v_s is the final velocity (equal to zero since the boulder comes to rest). Thus

$$u_s^2 = 2fx \tag{3.5}$$

But

$$u_s = v_f \sin \alpha \tag{from (3.4)}$$
$$v_f^2 \sin^2 \alpha = 2fx$$

But

$$v_f^2 = 2gh \tag{from (3.3)}$$
$$2gh \sin^2 \alpha = 2fx$$

Re-arranging terms gives:

$$h/x = f/(g \sin {}^2\alpha) \tag{3.6}$$

The boulder, say of mass m, generates on the scree slope a downslope force component of $mg \sin \alpha$ (same trigonometry as figure 3.6), and this is opposed by an uphill force resultant upon the mass of the boulder and the coefficient of sliding friction which can be represented as a tangent, say $\tan \phi_d$. The magnitude of this frictional force is obtained by multiplying the force of the boulder normal to the slope ($mg \cos \alpha$) by $\tan \phi_d$. The resultant force on the boulder is obtained by subtracting the upslope ($mg \cos \alpha \,.\, \tan \phi_d$) from the downslope ($mg \sin \alpha$) forces, i.e.

$$\text{resultant force} = mg \cos \alpha \,.\, \tan \phi_d - mg \sin \alpha \tag{3.7}$$

This will have a positive value if $\alpha < \tan \phi_d$. It will be recalled that a force is obtained by multiplying a mass by its acceleration, so the acceleration of a mass with known associated force can be determined by dividing the force by the mass. Thus the acceleration (upslope) of the boulder is found by dividing equation (3.7) by m. So

$$f = g \cos \alpha . \tan \phi \ - g \sin \alpha$$
$$= g(\cos \alpha . \tan \phi \ - \sin \alpha) \tag{3.8}$$

From equation (3.6)

$$f = (hg \sin {}^2\alpha)/x$$

Thus

$$(hg \sin {}^2\alpha)/x = g(\cos \alpha \,.\, \tan \phi_d - \sin \alpha)$$
$$h/x = (\cos \alpha \,.\, \tan \phi_d - \sin \alpha)/\sin {}^2\alpha \tag{3.9}$$

This approach necessarily makes quite a number of assumptions about the nature of rockfall and scree forming processes, but equation (3.9) can be tested as an appropriate model by comparing predicted results with those from field and laboratory experiments. Statham (1976) carries out such exercises and concludes that the theoretically derived model at least provides a satisfactory first approximation to account for particle movement on scree slopes.

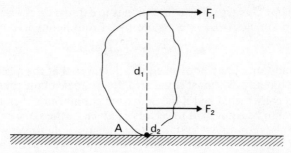

Fig. 3.11 Illustration of two different moments ($F_1 d_1$ and $F_2 d_2$) which can be applied to topple the boulder about point A.

3.3 Moment of a force, equilibrium, centre of gravity

The effect of a force on a body depends on the line of action of the force. This principle should be evident from many aspects of everyday life—it is easier to turn a stiff nut with a long spanner than with a short one. Consider a large boulder resting on a horizontal surface as shown in figure 3.11. The easier strategy to push this boulder over is to exert a force at the top of the boulder (F_1) rather than lower down (F_2). The boulder rotates about point A. The product of the magnitude of a force and its perpendicular distance from the axis of movement is known as the *moment*. In figure 3.11 the moment of force F_1 is $F_1 d_1$ whilst for F_2 the moment is $F_2 d_2$. It should thus be clear why F_1 is a more effective force in moving the block compared to F_2, even though the magnitudes of F_1 and F_2 may be the same. The important principle to remember is that the length of the moment arm is the perpendicular distance from the line of action of the force to the axis of movement. Figure 3.12 shows in plan view an irregularly shaped slice of wood; imagine that various forces are exerted on the perimeter of this slice as indicated on the figure.

The problem is to determine the magnitude of the moments about point O where a state of equilibrium exists. There will be no movement if the sum of the

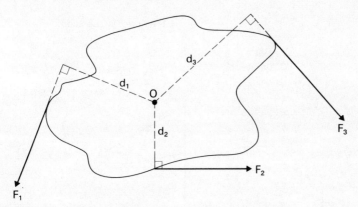

Fig. 3.12 Three forces (F_1, F_2 and F_3) are applied to the perimeter of an irregularly shaped slice of wood. At equilibrium, the sum of the moments about point O equals zero.

moments about O is zero. The convention is to express clockwise moments as negative and anticlockwise ones as positive. From figure 3.12

$$F_1 d_1 + F_2 d_2 - F_3 d_3 = 0$$

This is a condition of any body in equilibrium, viz. that the algebraic sum of moments about any axis must equal zero. If a triangle of the three vectors F_1, F_2 and F_3 is constructed and if the body is in equilibrium, then the vector F_3 can be shown to be equivalent to vectors F_1 and F_2—the same argument as in figure 3.5. This is equivalent to stating that the sum of the forces when resolved along both the X- and Y-axes must each equal zero when a body is in equilibrium.

Various types of equilibrium are possible, but these can only be explained after brief mention has been made of centre of gravity. The centre of gravity of a body is the point within the body at which all the mass can be considered to act. If the body is a cube and is made of homogeneous material, then the centre of gravity will be at the point in the centre equidistant from all six sides. With irregularly shaped bodies, the centre of gravity can be determined by empirical techniques or by integration; this latter method will not be discussed. Instead the concept can be amplified by considering a thin irregularly shaped body which is suspended (figure 3.13). A plumb-bob is also attached to the first point of suspension and thus a vertical line can be drawn on the body when it has come to rest. The procedure can be repeated at other points as shown in the diagram and all the lines will intersect at the centre of gravity (O). The same approach can be adopted in theory for bodies which are irregular in three dimensions.

A return can now be made to a consideration of equilibrium. Figure 3.14 (a) shows a solid cube of mass m resting on a horizontal surface. The dot indicates the centre of gravity. A force F_1 is applied to the upper corner as shown in figure 3.14 (b) so that the cube is tilted up and swivels about point A. If the force is kept constant to maintain the cube in the position as illustrated in figure 3.14 (b), then the moment Fd_1 must equal mgd_2. Now if the disturbing moment Fd_1 is removed, then the effect of the restoring moment mgd_2 will be to return the cube quickly to its original position. This is a demonstration of stable

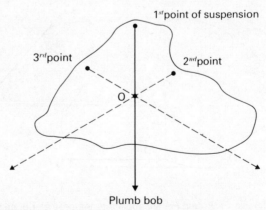

Fig. 3.13 The suspension of an irregularly shaped slice of wood from three points in order to determine the centre of gravity (O) of the body.

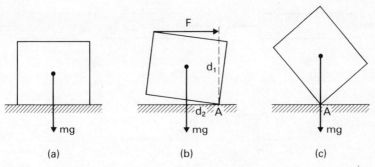

Fig. 3.14 A cube of mass *m* rests on a horizontal surface (a). In (b) a force has been applied to tilt the cube about A. On release of *F*, the cube reverts to the initial position, whilst in (c), the cube can easily topple away from its initial position.

equilibrium—if disturbing forces are removed, then the body will revert to its initial situation. As the cube is more and more raised on end by force *F*, so the magnitude of the restoring moment will decrease until the situation in figure 3.14 (c) is achieved whereby this latter moment is zero. Necessarily the applied force *F* must also be zero in this position. In effect the cube is balanced in a very delicate state—unstable equilibrium or metastable equilibrium. A very slight displacement to the left would cause the restoring movement to achieve situation (a), but a similar movement to the right would mean that the moment mgd_2 would quickly increase in magnitude pulling the cube over into a new position and this could occur without the assistance of the external force *F*. In other words there would be an acceleration away towards a new state of equilibrium. The relevance of these principles can be illustrated by brief mention of gully formation and metastable equilibrium, a theme which has been developed by Hudson (1971). The interpretation is that a state of metastable equilibrium can arise on a hillslope and then gullies can start and quickly accelerate in development once there is slight disturbance in the equilibrium state. The gullies will develop quickly, completely change the initial form of the slope and eventually will result in a new equilibrium situation. The clear implication of this view of gully development is that it is misleading to try to relate the initiation of individual gullies to specific factors, but instead concern should be with the overall equilibrium state on the hillslope and its change.

3.4 Stress and pressure

Soil creep or soil slippage result because downslope force components cannot always be matched by stronger forces of soil resistance. The science of soil mechanics is concerned with the behaviour of soil to applied forces—a theme directly relevant to geomorphic processes. In order to consider some fundamentals of soil mechanics, certain vectors derived from forces must be introduced. Consider a horizontal rod which is subjected to a force at either end as in figure 3.15 (a). The rod is under tension and it will only fracture at right-angles to the force if this cross section cannot withstand the force per unit area. *Stress* is the vector obtained by dividing the force by the cross-sectional

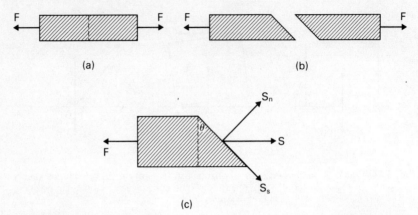

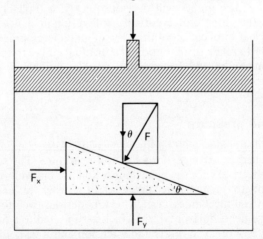

Fig. 3.15 (a) A bar under tension through the application of force *F*. In (b) fracture has taken place along a plane at an inclined angle and the associated stresses are shown in (c). S_n is the normal stress whilst S_s is the shearing stress. (after Sears and Zemansky 1963, p. 252)

area. In situation (b) fracture has taken place such that the fracture plane is not normal to the direction of the force; with this fracture the cross-sectional area is greater than in (a). The stress vector (S) is indicated in (c), and because it is a vector it can be resolved into two components, one at right-angles to the fracture surface (S_n) and the other parallel to the fracture surface (S_s). S_n is called the *normal stress* because it is the component of stress normal to the fracture surface whilst the other stress S_s is the *shearing stress*; it is this stress component that will lead to slippage along the fracture surface.

Imagine a small volume of soil within a large confined mass and consider this small volume to be of cube shape. The six sides of the cube will be subjected to normal and shearing stresses. The normal stresses will necessarily be at right-angles to the sides of the cube. If a three-dimensional coordinate system is selected such that the three axes are parallel to the sides of the cube, then the

Fig. 3.16 The shaded triangle represents a wedge-shaped part of a fluid under hydrostatic pressure. Forces F_x, F_y and F are the forces normal to the three sides of the triangle. (after Sears and Zemansky 1963, p. 254)

three normal stresses can be represented as σ_x, σ_y and σ_z which can be resolved into one normal stress, σ. Similarly the shearing stresses on each face can be resolved into one shearing stress, τ. The convention in soil mechanics is to use the Greek letter σ (sigma) for normal stress and τ (tau) for shearing stress. For simplicity the reference frame was selected by the orientation of the cube faces. It is possible to select a whole suite of reference frames and for each, normal and shearing stresses could be obtained by resolution of vectors. However it can be found by trial and error, or by theoretical arguments, that there exists one particular orientation of a three-dimensional coordinate system such that the shear stress vector disappears or becomes zero. Thus there are three characteristic planes, called *principal planes*, all at right-angles to each other within the soil mass, upon which there are only normal stresses. The axes for the frame are known as *principal stress axes*. The principal planes are ordered in terms of their magnitude: major (σ_1), intermediate (σ_2) and minor (σ_3).

It should be clear from dimensional considerations that stress is equivalent to pressure and this can be demonstrated by considering either a gas or a liquid under pressure as shown in figure 3.16. Consider a wedge-shaped part of the medium and assume that there is no movement within the container implying that shearing stresses are non-existent. For simplicity neglect the mass of the gas or liquid and thus the only forces which are exerted on the wedge are at right-angles to the sides, F, F_x and F_y as in figure 3.16. A triangle of forces can be constructed as indicated to show that

$$F \sin \theta = F_x \text{ and } F \cos \theta = F_y$$

Let A, A_x and A_y correspond to the areas of the sides of the wedge as illustrated and thus

$$A \sin \theta = A_x \text{ and } A \cos \theta = A_y$$

When the upper equations are divided by the lower,

$$\frac{F}{A} = \frac{F_x}{A_x} = \frac{F_y}{A_y}$$

This suggests that the value of force per unit area is the same in any direction and these ratios define the *hydrostatic pressure p* in a liquid or gas. Hence

$$p = \frac{F}{A} \text{ or } F = pA$$

Pressure has dimensions resultant upon dividing a force by an area, for example newtons per square metre or pascals. In meteorology, pressure can be expressed in terms of the height of mercury which the atmosphere supports in a tube. When a tube full of mercury and closed at one end is inverted in a bowl of mercury, the level of the mercury drops to a height of about 0.76 m above the level in the bowl. This height of mercury corresponds to an atmospheric pressure of 1.013×10^5 N m^{-2}, sometimes called an atmosphere, a term which will be used for particular situations later in the book. The measure, bar, is often used in meteorology and 1 bar is taken to be equivalent to a pressure of 10^5 N m^{-2}; because 1 bar is a large unit it is subdivided into millibars and these are the units which are often used on meteorological charts to show variation in pressure (see appendix 2 for pressure conversions).

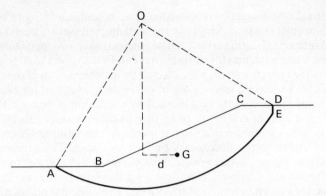

Fig. 3.17 Cross section of an area liable to a rotational landslide with O the centre of the circle, AE the potential slip plane and G the centroid of area ABCDE.

Brief analysis of a landslip provides an opportunity to illustrate some of the principles described in sections 3.1 to 3.4. A landslip results from failure in soil, meaning that one mass of soil moves suddenly with respect to another; the surface between these masses is called the slip plane. Figure 3.17 shows the original cross section of a slope in which a landslip has occurred. The slip plane is defined by the arc AE and the centre of rotation of the soil mass is at O. DE is a vertical crack at the top of the landslip. Imagine that the figure shows a slice of the soil which slipped and this slice is of unit thickness. The need is to understand the circumstances which led to the landslip and this is possible through a consideration of moments. The tendency for the soil mass to slip is resultant upon the disturbing moment whilst the forces exerted by the soil along the potential slip plane cause a resisting moment: these two types of moment require some explanation. If a pendulum is held away from the vertical, the moment of the pendulum can be obtained by multiplying the force resulting from its mass by the distance from a vertical line through the point of suspension; the further this distance, the greater will the moment be. Exactly the same approach is applicable to the slice of soil which slides along the arc EA. The centre of gravity (G) of this slice has to be determined following the method as demonstrated in figure 3.12. This point G is analogous to the pendulum. The soil rotates about O and thus the disturbing moment is obtained by multiplying the mass of the unit slice in area ABCDE by the acceleration due to gravity by the horizontal distance of G from the vertical line through O. The resisting moment depends upon the length of the slip plane (AE), a measure of soil cohesion (see chapter 7 for a fuller description of this property), and the distance of the slip plane from the point of rotation (the radius of the arc). Thus the two moments can be expressed as follows:

$$\text{disturbing moment} = V\gamma g d$$
$$\text{resisting moment} \quad = acr$$

where

      V is the volume of slice ABCDE with unit thickness
      γ is the bulk density of soil
      g is the acceleration due to gravity

d is the horizontal distance as in figure 3.17
a is the area of arc AE which is of unit width
c is the cohesion of soil
r is the distance from 0 to AE

Three situations are possible:
(1) Disturbing moment < resisting moment.
    Under these circumstances, no slippage can take place.
(2) Disturbing moment = resisting moment.
    This is a highly unstable circumstance akin to the situation as presented in figure 3.14 (e)—metastable equilibrium.
(3) Disturbing moment > resisting moment.
    Slippage results from this situation.

In a slope stability analysis then, the critical characteristics are the relative magnitudes of the disturbing and resisting moments. The ratio of the resisting to the disturbing moment is known as the safety factor. The design of slopes along new motorways, for example, should ensure that predicted values of the safety factor should be well in excess of 1·0. In practice it should be noted that the value of the safety factor will vary not only according to possible slip planes, but also according to variations in the values of the variables included in the moment computations.

3.5 Work, power and energy

The previous section showed how stress and pressure can be derived by dividing force by area. Work is also derived from force in that the expenditure of work implies the application of force over a distance. In order for a stream to move a boulder from one position to another, an adequate force must be exerted over the required distance. In a strict sense when a force is applied to move a body, the work done by the force is defined as the product of the displacement and the component of the force in the direction of the movement. Consider the situation illustrated in figure 3.18. The force F is not being applied in the direction of the movement and thus the work done over distance x is $(F \cos \theta) x$. In practice the magnitude of the force often varies during the time when work is being expended. In such a situation it is necessary to sum up all the component work values from the initial (x_1) to final position (x_2). This can be expressed in integral form:

$$W = \int dW = \int_{x_1}^{x_2} F \cos \theta \, dx$$

The unit of work is the *joule* which is the work done when a force of one newton acts over a distance of one metre.

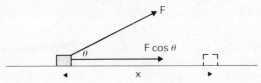

Fig. 3.18 The application of a force F at an angle θ to move the block over the distance x.

Power is the rate of doing work and thus is obtained by dividing the work done by the necessary time period. In order to push a body from one point to another a small force can be exerted over a long period of time or a large force over a short period of time; the net effect is the same, but with the second strategy higher power must be developed. The unit of power is the *watt* and this is achieved when 1 joule is being expended per second. Man can only develop about 50 W when working over long periods, but short surges of up to 250 W are possible (Duncan 1975).

Energy is dimensionally equivalent to work and the SI unit of energy is the joule (see appendix 2 for other units). Energy can be defined simply as the ability or capacity to do work. Waves crashing against a cliff expend energy; the process of photosynthesis in leaves depends upon the receipt of energy from the sun; the formation of a meander demands the expenditure of energy. In fact all processes within the physical environment involve the transformation of energy. The following chapter examines the principles of energy in far greater detail, but it is appropriate to conclude this chapter with an outline of the two types of mechanical energy—*potential energy* and *kinetic energy*.

The potential energy of a body results from its position relative to some reference level. A body of mass m held at a height h above a ground surface possesses a quantity of potential energy since in falling from its initial point to the ground it can perform work. In fact, the potential energy is mgh where g is the acceleration due to gravity. As the body approaches the ground, so the potential energy decreases to become zero at the ground. Thus the potential energy of a body is due to its position.

In contrast the kinetic energy of a body results from its motion. A body of mass m, travelling at velocity v, can perform work in being brought to rest; in fact the value for kinetic energy is obtained from calculating $\frac{1}{2}mv^2$. A free-falling body, assuming there is no air resistance, accelerates at g, thus the potential energy decreases as the body approaches the ground whilst the kinetic energy increases correspondingly. The practical importance of potential and kinetic energy can be hinted at by consideration of the longitudinal profile of a river as shown in figure 3.19. At locality A, the tip of the channel, the mechanical energy of a hypothetical unit volume of water could be calculated by adding the potential and kinetic energies. The former is obtained by multiplying the mass of the volume by g and h_1, the height above base level. The kinetic energy is derived by multiplying half the mass of the unit

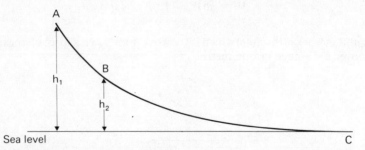

Fig. 3.19 The curve represents the longitudinal profile of a river from source (A) to the sea (C). The elevations at points A and B are indicated.

volume by the square of the velocity. At point B the potential energy is less than at point A because of the drop in altitude whilst the kinetic energy is greater than at A because there is the tendency for the average velocity of flow in channels to increase downstream. At point C, the mouth of the river, the potential energy of a unit volume is zero. Thus the potential energy of constituent unit volumes of water decreases along the length of the stream. In contrast the kinetic energy of the same parcels of water will vary according to the square of the velocity: the tendency in streams is for velocity to remain constant or to increase downstream (Leopold, Wolman and Miller 1964) and the kinetic energy must therefore stay the same or increase. The variation in the sum of kinetic and potential energies of unit volumes along the length of a channel raises many interesting problems. Does the increase in kinetic energy when it occurs compensate for the decrease in potential energy? If not, how does the channel respond in order to increase its loss of energy? Can the equilibrium profile of a river be understood with reference to energy principles? These questions indicate that an appreciation of energy principles is fundamental to fluvial geomorphology and the same argument could be presented for many systems in physical geography. It is thus clear that attention must turn in the next chapter to the principles of energy since their application is fundamental to an understanding of the physical environment.

4
Principles of energy

Every process in our physical environment requires energy for its maintenance and thus it is vital that some detailed consideration be given to the principles of energy. As an example Bloom (1969) portrays in graphic terms the need for an energy approach to geomorphology. His geomorphic machine (figure 4.1) requires coal as a fuel so that it can perform all the grinding, sawing and filing of the landscape. The rates of these processes depend upon the energy output of the machine which depends in turn on the fuel input and the machine efficiency at converting coal to mechanical energy. The transference of energy is also illustrated in this example and thus the established laws concerned with the conservation and transference of energy must be discussed; these are called thermodynamic laws. Processes in the physical environment, be they physical or chemical or some combination, ultimately tend towards a state of

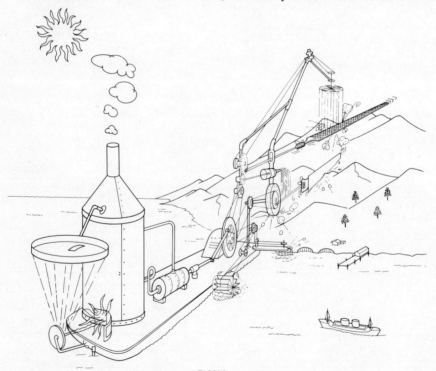

Fig. 4.1 The 'geomorphology machine' (from Bloom 1969, p. 9)

equilibrium if time permits. In such situations, functional parts of the environment, systems, reach stability reflected in constancy of form. Examples are specific soil profiles, hillslope profiles or vegetation complexes which do not continue to change in type or form. Such a stage in each case is characterized by specific energy distributions and thus the migration of systems to equilibrium states can be approached by applying the principles of energy distribution. If a key chapter has to be identified in this book, perhaps this is the one, such is the importance of energy principles.

The science of *thermodynamics* may sound rather awesome, but in outline it is concerned with the relationships between forms of energy, in particular heat and work. The science can be traced back to the early days of the industrial revolution when it became essential to understand the principles governing the conversion of heat into mechanical energy (Morse 1969). Such work predated the research on atomic structure towards the end of the nineteenth century when not only was the internal structure of atoms considered, but also the interactions between atoms were investigated; such an approach led to a deeper understanding of properties like the pressure and temperature of gases. This provided the basis to *kinetic theory* which explains the nature of gas pressures in part due to the kinetic energy of the constituent molecules. Fundamental research was also carried out on the statistical relationship between the atomic structure of any element and its microscopic behaviour, a topic known as *statistical mechanics*. Thus in this brief foray into the nature of energy, the interrelation of principles drawn from thermodynamics, kinetic theory and statistical mechanics will be shown.

4.1 The nature of systems and the first law of thermodynamics

An energy approach first demands that the system be defined: there can be no analysis unless the system is specified as a particular grouping of interrelated components which are separated from surrounding conditions. The search is usually for systems which display internal functional linkages; for example in hydrology or geomorphology a convenient unit is a drainage basin. The scale of analysis depends on the problem being investigated; in biogeography ecosystems can vary from the few square metres of a particular vegetational–faunal community to thousands of square kilometres of ecosystems such as tropical rainforest. Systems within the physical environment can never be considered as completely isolated since there are always some linkages across their boundaries. Certain properties characterize any system; for example a drainage basin system is defined in terms of the area of land within which surface runoff and subsurface water flows towards a particular stream or stream network. In thermodynamics, such properties as temperature, pressure, volume and chemical composition specify the nature of a system and are called *state variables*. If the system is at equilibrium, then the values of these variables are characteristic of the specific state of the system and in no way indicate the previous history of the system.

A useful way to introduce some thermodynamic principles is to consider a system composed of a gas enclosed in a cylinder with a movable piston. Suppose there is a flow of heat (Q) from the surroundings, through the cylinder to the gas which will then expand; the result is that the piston will be moved and work performed (W). As will be discussed not all the heat added to the system is transformed into work done by the piston. This means that some of the thermal energy is added to the *internal energy* (E) of the system. The ability of a gas to retain energy is the result of the energies possessed by individual molecules—kinetic, potential, rotational and vibrational energies (chapter 7). If E_1 is the initial internal energy of the system and E_2 is the internal energy after the inflow of heat, then the increase in internal energy is $(E_2 - E_1)$, usually represented as ΔE. This internal gain in energy must equal the difference in energy between that absorbed by the system (Q) and the work performed by the system (W). Thus

$$\Delta E = Q - W \qquad (4.1)$$

This equation is a mathematical statement of the *First Law of Thermodynamics* and expresses conservation of energy. Equation (4.1) can be written as

$$Q = \Delta E + W \qquad (4.2)$$

This states that the energy added to a system equals the sum of the increase in internal energy and the energy expended by the system. Certain sign conventions are implicit in equations (4.1) and (4.2). The numerical value of Q is positive when heat is absorbed by the system and negative when it measures heat evolved by the system. The value of W is positive when work is done by the system, but is negative when work is done on the system.

The first law of thermodynamics is of direct application to a number of situations in physical geography. The conservation of energy in ecosystems will be demonstrated later in this chapter and the nature of energy budgets will be further illustrated in chapter 6. Closely allied to the law of conservation of energy is the law of conservation of matter. Special problems arise with this law when nuclear changes occur with associated release of energy, but such difficulties can be resolved given Einstein's linkage of matter and energy. There are many situations in physical geography where a mass budgeting approach is particularly useful. The obvious example to quote is the hydrological cycle for a drainage basin: the water budget can be expressed in a similar way to equation (4.2) as follows:

precipitation = changes in storage + basin channel runoff +
　　　　　　　　evapotranspiration

$$P = \Delta(I, R, M, L, G, S) + q + e \qquad (4.3)$$

where

> P is precipitation
> I is interception storage
> R is surface storage
> M is soil moisture storage
> L is aeration zone storage
> G is ground water storage
> S is channel storage
> q is basin channel runoff
> e is evapotranspiration
> (from More 1969).

Another application of this conservation of matter is in hillslope geomorphology through the use of continuity equations. Equation (4.3) is a

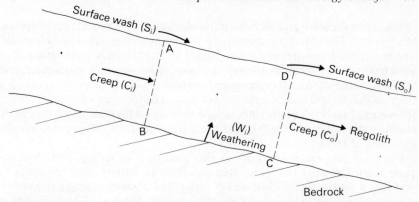

Fig. 4.2 Cross section of a volume of regolith (ABCD) on a slope; inputs and outputs to ABCD due to surface wash and creep are indicated as well as weathering input.

continuity equation expressing conservation of matter, but others can be developed. Figure 4.2 is a cross section of part of a hillslope mantled with regolith which overlies bedrock. The unit to be considered consists of the volume of regolith, indicated in the cross section by ABCD. A budget approach requires an accounting of the inputs and outputs to this volume. Suppose the geomorphological processes can be limited to creep, surface wash and weathering of the bedrock. Let the various processes be represented as follows:

c_i : creep input
c_o : creep output
s_i : surface wash input
s_o : surface wash output
w_i : weathering input

These measures can be expressed in terms of volume per time unit. Then the change in volume (ΔV) over this time period can be determined as follows:

$$(c_i + s_i + w_i) - (c_o + s_o) = \Delta V \qquad (4.4)$$

This statement is an example of a continuity equation and from it could be derived others to express change in surface elevation and soil thickness in terms of the inputs and outputs. More sophisticated statements of continuity require the use of calculus, an approach developed by Carson and Kirkby (1972). The great merit of such an approach is in solving these continuity equations when selected processes such as creep or wash can be expressed in terms of slope angle or distance from hillslope crest. This means that continuity equations allow the prediction of hillslope form on the basis of particular processes (figure 1.1).

4.2 Principles associated with energy transfer: the second law of thermodynamics

Brief examination of equation (4.1) will reveal that an infinite number of values of Q and W could be selected always giving the same value for ΔE. The input of heat (Q) results in a change of state of the system; before this input, values for pressure, volume and temperature (state variables) within the cylinder characterize the state and after the input new values for the state variables will be apparent. The time, for example, over which the change takes place from the first to second state is irrelevant with regard to the value of ΔE. The term *path* is used to describe the nature of system change from one state to another, and thus the value of ΔE is independent of path; such a quantity is called a *thermodynamic function* or a *function of state*. The measure of internal energy (E) is also a function of state since it depends only on the state of the system. In contrast the input of heat (Q) and the output of work (W) are not usually functions of state.

Chemical processes can lead to the liberation of energy and thus the performance of work, but it is often necessary to initiate such processes by applying a short concentrated energy input. The obvious example to quote is the petrol engine, where the reaction between the petrol and oxygen is activated by the spark from the plug. This input of *activation energy* raises the energy level of the constituent molecules to an unstable condition whence they react to stabilize after the release of energy at a lower energy state, a process which is represented in figure 4.3 (a). This reaction is *exothermic* since heat is given off and the activation energy is more than returned. In contrast, an *endothermic* reaction absorbs heat, and a suitable example from the physical environment is photosynthesis (figure 4.3 (b)). In this case the activation energy is provided by solar radiation which results in the production of sugar and oxygen. The net energy output is less than the activation energy. In fact, every chemical reaction involves an energy change; just as a chemical equation expresses quantitatively the nature of a reaction, so a thermochemical equation also describes the magnitude and direction of heat flow. As an example, consider the ignition in a bunsen burner of one mole of methane (CH_4) with two moles of oxygen represented by the following thermochemical equation:

$$CH_4(g) + 2O_2(g) \rightarrow CO_2(g) + 2H_2O(l) : Q = -890 \cdot 4 \, kJ$$

The symbol Q indicates heat flow; the convention as already mentioned is to

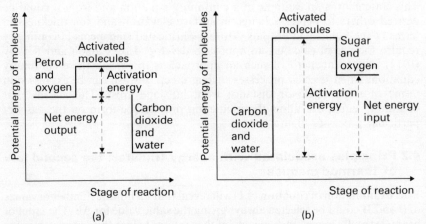

Fig. 4.3 The change in potential energy of molecules during a reaction. (a) illustrates an exothermic reaction whilst (b) shows an endothermic one. (after Weaver 1972, p. 9)

use a negative sign if the reaction is exothermic and a positive sign if it is an endothermic one. It is important to note that the value of Q depends on the states of the reactants and products and that a uniform temperature of 25 °C is assumed. It would be possible to ignite the same volumes of methane and oxygen in a robust closed cylinder and measure the amount of evolved heat; this would prove to be slightly less, $Q = -885 \cdot 3 \, \text{kJ}$. The difference between the two situations is that ignition in the burner is at atmospheric pressure whilst this is not the case with the enclosed cylinder. In the burner situation, the value of Q is larger because work is done on the system by the atmosphere. In the physical environment chemical reactions at or very near the surface of the earth take place under atmospheric pressure, but with increasing distances either up into the atmosphere or down into the earth, the quantities of heat involved will change. For comparison purposes, values of heat must be calculated for constant-pressure conditions. The heat liberated or absorbed in such processes is called *enthalpy change*, but a fuller description of enthalpy is only possible once consideration is given to the cylinder system.

Suppose that the gas enclosed within the cylinder absorbs a quantity of heat (Q); as a result the gas expands and the piston moves from position 1 to position 2 (figure 4.4). Let the initial volume of the gas be V_1 and the final volume V_2 so that the volume change is $(V_2 - V_1)$. Suppose the corresponding change in internal energy is $(E_2 - E_1)$ or ΔE.

$$\text{work} = \text{force} \times \text{distance}$$

and

$$\text{pressure} = \text{force/area}$$

$$\therefore \text{force} = \text{pressure} \times \text{area}$$

Thus

$$\text{work} = \text{pressure} \times \text{area} \times \text{distance}$$

$$= \text{pressure} \times \text{volume}$$

The work (W) performed by the piston is moving from position 1 to 2 is as follows:

$$W = P(V_2 - V_1)$$

where P is pressure and is constant. From the first law of thermodynamics (equation (4.2))

$$Q = \Delta E + W$$

Thus

$$Q = E_2 - E_1 + P(V_2 - V_1)$$

$$= (E_2 + PV_2) - (E_1 + PV_1)$$

Let H_2 represent $(E_2 + PV_2)$ and H_1 represent $(E_1 + PV_1)$. Thus

$$Q = H_2 - H_1$$

$$= \Delta H$$

The quantity $(E + PV)$ is called the *enthalpy* or *heat content* of a system. It is the sum of the internal energy and the pressure–volume product. Since internal energy cannot be measured on an absolute scale, an arbitrary scale for enthalpy has also to be used. These values for enthalpy are of particular practical importance to refrigerators and steam engines; of greater relevance

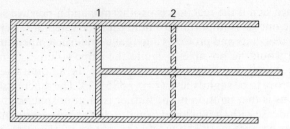

Fig. 4.4 A piston in a cylinder which moves from 1 to 2 as a result of expansion of the enclosed gas.

to processes in the physical environment is the quantity ΔH or *enthalpy change*, also called the *heat of reaction*. This equals the amount of heat absorbed in a process at constant pressure when the only work done is obtained by multiplying the pressure by the change in volume. Enthalpy change is a function of state, like internal energy. If the value of ΔH is negative then the reaction is exothermic whilst it is endothermic if ΔH is positive. If ΔH equals zero, then the reaction is isothermal. Enthalpy change implies a difference in enthalpy (H) between reactants and products.

Table 4.1 Selected heats of formation (ΔH_f°) and free energies of formation (ΔG_f°) at 25°C and one atmosphere pressure. l, liquid state; g, gaseous state; s, solid state; (adapted from Weast 1974).

Chemical compound	Heat of formation, ΔH_f° (kJ mol^{-1})	Free energy of formation, ΔG_f° (kJ mol^{-1})
Al_2O_3(s)	−1699·8	−1576·41
CaO(s)	−635·5	−604·17
$Ca(OH)_2$(s)	−986·6	−896·76
$CaSO_2$(s)	−1432·6	−1320·30
CO(g)	−110·5	−137·28
CO_2(g)	−393·5	−394·38
CuO(s)	−155·2	−127·19
Cu_2O(s)	−166·7	−146·36
CuS(s)	−48·5	−48·95
$CuSO_2$(s)	−769·9	−661·91
Fe_2O_3(s)	−822·2	−740·99
Fe_3O_4(s)	−1117·1	−1014·20
H_2O(g)	−241·8	−228·61
H_2O(l)	−285·9	−237·19
H_2S(g)	−20·1	−33·02
H_2SO_4(l)	−811·3	−690·10
KCl(s)	−435·9	−408·32
$MgCO_3$(s)	−1112·9	−1029·26
NaOH(s)	−426·7	−381·00
NH_3(g)	−46·2	−16·64
NO(g)	+90·37	+86·69
PbO(s)	−219·24	−189·33
PbO_2(s)	−276·6	−219·00
Pb_3O_4(s)	−734·7	−617·56
$SnCl_4$(l)	−545·2	−474·05
SO_2(g)	−296·9	−300·37

It is possible to determine ΔH experimentally by using calorimeters. If H was known both for the reactants and the products this would not be necessary but these values cannot be obtained. Instead the problem can be tackled if the values of a property called *heat of formation* (ΔH_f°) are known; this is the heat required for the formation of compounds from elements, the compounds and the elements being in their usual states (gas, liquid, or solid—called the standard states) at 25 °C and at one atmosphere pressure (Anderson, Ford and Kennedy 1973, p. 168). Values of ΔH_f° have been determined for a large number of compounds and a few are presented in table 4.1; a full list can be obtained from the *Handbook of Chemistry and Physics* (1974) edited by Weast.

Enthalpy change (ΔH) for any reaction can be obtained by subtracting the sum of the heats of formation of the products from the sum of the heats of formation of the reactants, i.e.

$$\Delta H = \sum \Delta H_f^\circ \text{ (products)} - \sum \Delta H_f^\circ \text{ (reactants)}$$

An example is the energy value of food which is taken to be the enthalpy change when the food is oxidized to such substances as carbon dioxide, water and various nitrogen compounds. This oxidation is supposed to be similar to the processes which take place within the human body. The result is that various metabolic processes are fuelled and people are also able to perform work. The average adult needs a minimum daily intake of the order of 10 000 kJ and for interest, some of the energy values for selected foods are presented in table 4.2.

It should be noted that human life demands not only a certain intake of food energy, but also small quantities of nitrogen, sulphur, vitamins, various minerals and amino-acids.

In terms of ecosystems, human beings are at the end of the food chain; only through parasites or after burial is energy returned to the ecosystem. Energy is obtained through eating meat, fish and cereals and other foods in a variety of forms, but the initial input of energy to ecosystems is always ultimately the result of radiation fuelling photosynthetic processes. Central to human existence and to every form of life is the transfer of energy through systems—this transfer can be by chemical or mechanical processes. The first law of thermodynamics states that in any system the *totals* of energy must balance,

Table 4.2 Energy values of selected foods (after McCance and Widdowson 1960).

Food	Energy value (kJ g^{-1})
Grilled steak	12·7
Roast mutton	12·2
Pork leg roast	13·3
Fried herring	9·8
Strawberries	1·1
Boiled potatoes	3·6
Fresh milk	2·8
Eggs	6·8
White bread	10·2
Bitter draught ale	1·2

but in no way does the law imply direction to processes within systems; no indication is given of the *availability* of energy. The *Second Law of Thermodynamics* copes with such problems, but in order to describe and explain its meaning, an outline of some more basic energy concepts is required.

In the last chapter it was inferred that there are distinct energy levels associated with particular electronic configurations of individual atoms. An input of energy can elevate the energy status to cause characteristic radiation; when a stream of electrons is passed into a neon-filled tube, a distinctive radiation is produced—the principle of fluorescent tubes or sodium street lights. If a solution is injected into a flame in a flame photometer, the strength of particular colours can be measured in order to determine the concentrations of individual elements such as sodium or potassium; the flame causes the necessary increase in energy which results in the radiation. The implication is that the molecules of a liquid in a beaker individually possess a certain amount of energy and these energies can be elevated by an appropriate energy input. The total energy of the liquid can thus be viewed as the sum of the energies of the molecules. But how do molecules possess energy? Perhaps the two most important types are *bonding energy* and *thermal energy* since these are of greatest relevance to chemical processes. As explained in chapter 2, molecules result from the bonding of atoms—these bonds between atoms within molecules have energy values. With solids and liquids the bonds between molecules assume significance; this intermolecular energy results from van der Waals' forces (chapter 7). Thermal energy results from the translational movement, rotation and vibration of individual molecules. Before the full meaning of this statement is explained the total energy of a substance can be summarized as follows:

bonding energy + thermal energy + other energy forms
= total internal energy of substances

If in Britain a weather forecast suggested that the temperature on the following day was to be about 22 °C, then a warm day would be anticipated—in other words the everyday assumption is that temperature is a measure of warmth or coldness. But temperature must result from thermal energy and thus it is the rates of translational movement, rotation and vibration of individual molecules that condition temperature. Consider a solid at low temperature; the main component of total internal energy will be derived from bonding energy since the individual molecules will only possess a small vibration and perhaps some rotational movement. If heat is applied to the solid, then the thermal energy of the molecules will increase to cause eventually a *phase change* when the solid melts; this occurs when the vibrational motion becomes sufficiently strong to break some of the bonds and thus translational movement of molecules is possible. Continued application of heat will add further energy to the molecules so that eventually individual molecules possess sufficient energy to break all bonds with adjacent molecules and thus move about more or less as free agents; this reflects the phase change from liquid to gas. At very high temperatures it is possible for electrons to be removed by molecular collisions to thus produce charged particles—ions. This process is called *ionization*. The details of phase changes are developed in chapter 6. One point worthy of stress is that temperature is a measurement of the *transfer* of heat; the response of a thermometer on insertion into a beaker of hot water

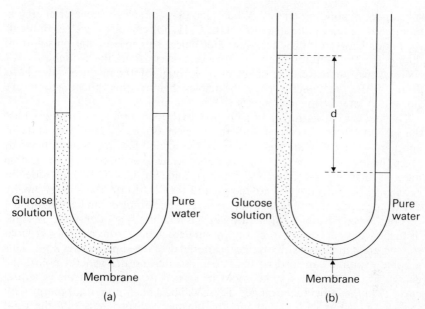

Fig. 4.5 Illustration of osmosis. (a) shows the initial situation and (b) the final one. *d* is the osmotic potential (exaggerated). (after Brady 1974, p. 172)

depends upon the transfer of energy into the thermometer causing the expansion of the mercury. Heat should not be confused with internal energy since the former is used to describe the transfer of energy.

It may be appropriate at this stage to demonstrate a particular phenomenon which can be explained by energy considerations. Figure 4.5 illustrates a U-tube, the limbs of which are separated by a semipermeable membrane which is a kind of preferential filter in that one substance may pass through it, but not another. All living cells are enclosed in such membranes; laboratory experiments requiring such membranes often use the skins of grapes. Into the right-hand limb is poured pure water and an equal volume of glucose solution is put into the other; these two liquids are separated by the membrane which allows water molecules through, but not the larger glucose molecules $(C_6H_{12}O_6)$. At first the situation in figure 4.5 (a) is observed, but after some time the liquids in the tube achieve levels as shown in figure 4.5 (b). Clearly there has been a movement of water molecules from right to left to build up the level of the glucose solution sufficiently to compensate for the movement of water through the membrane. This process is known as *osmosis* and the pressure indicated by d in figure 4.5 (b) is called the *osmotic potential*. The practical importance of osmosis is demonstrated by noting that since every living cell is encased by a semipermeable membrane, cell nutrition and stability are achieved by osmosis.

How can this phenomenon be explained in energy terms? Figure 4.5 suggests that the pure water is able to perform more work than the sugar solution since the solution is elevated to a higher level than the pure water. If the pure water can yield more work than the solution, then it must have more available energy. As explained, a key component of the total internal energy of substances is bonding energy. When glucose is dissolved in water energy is required to break up the structure of the $C_6H_{12}O_6$ crystals as well as to break the bonds between individual water molecules. This means that on solution, though the total amount of energy must remain constant, the *available* or *free* energy must decrease. The effect of dissolving solutes in pure water is to decrease the free energy. The principles behind this phenomenon are elucidated later in this chapter as well as the next.

The first law of thermodynamics expresses the conservation of energy whilst the second law is concerned with the transfer of energy from one state to another. The nature and significance of the second law are more difficult to grasp than the first, and thus it seems appropriate to begin by describing some relevant background concepts. Strahler and Strahler (1974) provide an excellent brief introduction to energy and they lead into the second law by describing a pile of oranges in a supermarket. The oranges can be built up in a very organized pyramid, a structure which is potentially very unstable. A slight disturbance is very likely to cause the sudden change from a highly ordered state to great disorder with oranges scattered over the supermarket floor. This graphic example can also be used to illustrate another general point; the energy of the 'orange system' could be viewed as the sum of the potential energies of the individual oranges. There will be a maximum total energy with the pyramid, but with the change to the more disordered state the total mechanical energy will drop to zero because it is most likely that every orange will end up on the floor, which is taken as the datum level. The loss in available energy between these two situations can be explained by frictional losses. The principle which emerges from this example is that a change in a system is marked by an increase in disorder and decrease in available energy. Necessarily there is also a direction to the change since it is impossible to disturb one orange on the floor when the highly disordered state has occurred in the hope of inducing the organized pyramid. The direction of change in a system is dictated by the tendency of the system to achieve a lower energy state and also to reach a more random configuration.

These preliminaries are sufficient to introduce the meaning of *entropy* which can be approached from either statistical mechanics or thermodynamics; a start may be made with the former leading on to the latter. Entropy is a measure of the degree of disorder or randomness in a system. Several other aspects characterize the concept of entropy. The entropy for the whole of a system is the sum of the entropy of the parts of it and the entropy of a perfectly ordered system will be zero. Entropy can be produced but can never be destroyed. It is interesting to note the contrast between energy and entropy; the former can neither be produced nor destroyed, whilst the latter can only be produced. An expression which defines entropy is:

$$S = k \log_e \Omega \qquad (4.5)$$

where S represents entropy, k is a constant known as Boltzmann's constant and Ω is the number of possible configurations of the system. This latter term requires some explanation. If the orange example is recalled there are two states—either total order with a pyramid of oranges, or total disorder with oranges scattered over the floor. Imagine that toy building blocks are used instead of oranges and that it is possible to conceive of a variety of partially tumbled-down structures between total order and disorder. Each of these possible structures would have a particular total potential energy resultant upon the heights of the blocks above the ground. Further, probabilities could be assigned to these states to express the chances that they will occur (Ω). Reynolds (1974) takes a graphic example to demonstrate this concept; he considers a tray on which there are 100 red jumping beans and 100 white. There is only one way to have the tray all white on one side and all red on the other, so Ω is equal to 1 and hence the entropy value is zero in this perfectly ordered state. Then the beans begin to jump and gradually the whole tray will appear a uniform pink; there are many ways in which the beans could be distributed to achieve this. Necessarily Ω now must be a large number and entropy has correspondingly increased. It is hoped that this simple example demonstrates the statistical nature of entropy; as the number of possible internal configurations or dispositions increases, thus the randomness or disorder also increases resulting in an increase in entropy.

In a complex real world system the constituent states are likely to have different probabilities and thus equation (4.5) has to be written as:

$$S = -k\sum p_v \log p_v \tag{4.6}$$

where k is Boltzmann's constant and p_v is the probability of occurrence of the particular states $v_1, v_2, v_3, \ldots$. Equation (4.6) seems at first sight rather different from equation (4.5); consider the situation where all states are equally likely—the individual probabilities must equal $1/\Omega$ if there are Ω states. Thus:

$$S = -k\sum \frac{1}{\Omega} \log_e\left(\frac{1}{\Omega}\right)$$

$$= -k\left[\frac{1}{\Omega}\log_e\left(\frac{1}{\Omega}\right) + \frac{1}{\Omega}\log_e\left(\frac{1}{\Omega}\right) + \frac{1}{\Omega}\log_e\left(\frac{1}{\Omega}\right) + \ldots\right]$$

$$= -k\left[\frac{\Omega}{\Omega}\log_e\left(\frac{1}{\Omega}\right)\right] = -k\log_e\left(\frac{1}{\Omega}\right) = -k(\log_e 1 - \log_e \Omega)$$

$$= k\log_e \Omega$$

In order to relate this statistical approach of entropy to thermodynamics, some thermodynamic fundamentals must be introduced. An appropriate way to begin is to describe the classic example of the Carnot cycle in which there is a net heat flow into the system and net work is produced from the system, thus constituting an engine. Consider a piston in a perfectly insulated cylinder as in figure 4.6 and let the system be a gas in the volume between the piston and the cylinder. If a heat source at temperature T_h is applied at the end of the piston and heat (Q_h) is transferred to the gas, then the gas will expand and push the piston to the right (assume there is no frictional loss). Work is then performed by the system (W_E). The heat source can be removed and replaced by a heat sink at temperature T_c to cause a heat transference from the gas to the sink (Q_c). At the same time the gas is compressed by work being performed on the system (W_p). The amount of compression can ensure that the system is returned to the same temperature and pressure conditions as at the start of the cycle. The arrows in figure 4.6 indicate the direction of flow of heat, but if the flow is with reference to the system, Q_h is positive whilst Q_c is negative. Because

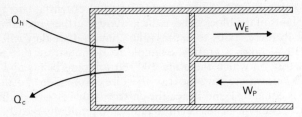

Fig. 4.6 A piston in a perfectly insulated cylinder to illustrate the nature of the Carnot cycle (see text for terms).

the system is ideal and because the initial and final situations are identical, the total change in energy (ΔE) for one cycle must be zero. According to the first law the sum of the heat flows must equal the sum of the expended effort either on or by the system:

$$\sum Q_{net} = Q_h - Q_c$$

and $$W_{net} = W_E - W_p$$

Also $$\Delta E = Q_h - Q_c - \sum W$$

But $$\Delta E = 0$$

Thus $$Q_h - Q_c = W_{net} \qquad (4.7)$$

As already mentioned, temperature is defined as being proportional to the amount of heat which is transferred, so T_h is proportional to Q_h and similarly T_c to Q_c. Thus

$$\frac{Q_h}{Q_c} = \frac{T_h}{T_c} \qquad (4.8)$$

In fact, this proportionality was taken by Kelvin as the basis for his absolute temperature scale. From equation (4.8)

$$Q_c = \frac{Q_h T_c}{T_h}$$

Efficiency must be a dimensionless value to express the ratio of net work produced divided by heat absorbed; it is given the symbol of η (eta). Thus from the above example,

$$\eta = \frac{W_{net}}{Q_h}$$

$$= \frac{Q_h - Q_c}{Q_h} \text{ from equation (4.7)}$$

$$= \frac{Q_h - (Q_h T_c / T_h)}{Q_h}$$

$$= \frac{1-(T_c/T_h)}{1} \text{(by dividing throughout by Q}_h\text{)}$$

$$= 1 - \frac{T_c}{T_h}$$

$$= \frac{T_h - T_c}{T_h} \tag{4.8}$$

This is an interesting result because it reveals two important points. The value of η would be 1 (maximum efficiency) if T_c was zero, a practical impossibility because 0 K is unobtainable; reduction in available energy is thus inevitable in a system. The value of η increases according to the difference between T_h and T_c, the temperatures of the heat source and heat sink. A practical demonstration of this from everyday life is the need for car engines to have effective cooling systems—heat sinks.

An underlying assumption for the Carnot cycle is that the processes are reversible, which means that at all stages of the cycle a process can be changed in direction. The notion of irreversible processes within a system has already been introduced with the pyramid of building blocks. It is impossible to get a car to accelerate by warming its brakes! These concepts require elaboration since, as will be demonstrated, the nature of change is incorporated into the second law. A system reaches equilibrium when state variables attain constancy of value. Strahler and Strahler (1974) demonstrate this by describing water from a tap running into a tank with a hole in its bottom. The level of water will build up until there is a sufficient head in the tank to cause the rate of outflow to equal the rate of inflow from the tap; this is only possible, of course, with a suitable flow from the tap and a hole of appropriate size. In this situation, water is constantly passing through 'the system', but dynamic equilibrium or steady state exists because the depth of water in the tank is constant. If the tap is turned on a little more, then the 'system' will pass through a transient state until a new constant depth is achieved—another dynamic equilibrium. This analogy can be applied to a wide variety of systems in the natural environment. Think of a soil system: a soil profile will have constancy of horizons and soil characteristics when the soil is in dynamic equilibrium even though the solid and liquid matter is passing down the system. With this interpretation, a very thin soil on a scree slope could be in dynamic equilibrium and such a view immediately invalidates the youthful and mature approach to soil evolution. With the tank example, and indeed with the soil system, any modification in input will lead to a *spontaneous change* associated with the transient state. It is possible that the tap could be slowly adjusted so that 'the system' always stays in equilibrium. These two situations illustrate the distinction between reversible and irreversible processes. If a transient state is established through a sudden turn of the tap, the reaction of the system is spontaneous meaning that the process is irreversible—the resultant change only goes in one direction. In contrast if 'the tank system' is always at equilibrium despite variations in inflow, then the process is always reversible. This implies that the system can respond in either direction at any time in order to maintain equilibrium. Reversible processes are indicated by a double

(*

arrow. For example imagine a beaker of water completely enclosed within a bell jar. Then there is an exchange of water from the liquid to gaseous state and vice versa:

$$H_2O(l) \rightleftharpoons H_2O(g)$$

When a state of equilibrium is achieved, the rate of evaporation will be exactly matched by the rate of condensation. Many chemical reactions are reversible and can be represented in this form with reactants and products separated by a double arrow. For the maintenance of such processes, the reactants and products obviously must be kept in contact.

A stage has now been reached whereby it is possible to consider the thermodynamic nature of entropy. Equation (4.8) can be rewritten as

$$\frac{Q_h}{T_h} = \frac{Q_c}{T_c}$$

or

$$\frac{Q_h}{T_h} - \frac{Q_c}{T_c} = 0$$

This ratio of Q/T is the basis of entropy (S). Imagine the cylinder and piston situation again. An input of heat (Q) results in the system performing work (W) and the change in internal energy (ΔE) is the difference between Q and W. ΔE is a function of state. Entropy, or rather change in entropy, can be approached in a similar way. Suppose the cylinder and piston system can change in a variety of ways or paths from its first to its second state and that these paths are always reversible. This is simply stating that it is possible to apply Q in a variety of ways, but with the same end result. Any path between the two states can be subdivided into a large number of heat inputs which sum to Q; if all these heat components are divided by the corresponding absolute temperature of the system and then added for the whole path, the result is the same for all paths between the two states as long as the paths are reversible. The value of this sum from state 1 to state 2 can be expressed as

$$\int_1^2 \frac{dQ}{T}$$

This function is the entropy change of the system (ΔS) between the two states. Thus (Sears and Zemansky 1963)

$$\Delta S = \int_1^2 \frac{dQ}{T} \tag{4.9}$$

The units of entropy must be energy per temperature: in the SI system joules per Kelvin degree $(J\ K^{-1})$. A good example to demonstrate a calculation of entropy is the melting of ice, say 1 kg at 0 °C. Now

$$\Delta S = \int \frac{dQ}{T}$$

The change takes place at constant temperature and thus the equation can be rewritten as

$$\Delta S = \frac{1}{T}\int dQ$$

$$= \frac{Q}{T}$$

To melt 1 kg of ice at 0 °C, 333·7 kJ are required, and thus by remembering that T always refers to absolute temperature, ΔS is calculated as follows:

$$S = \frac{333 \cdot 7}{273 \cdot 15} \text{ kJ K}^{-1}$$

$$= 1 \cdot 22 \text{ kJ K}^{-1}$$

This example shows that in an isothermal process the value of ΔS is found by dividing the applied heat (Q) by the absolute temperature (T), i.e.

$$\Delta S = \frac{Q}{T} \text{ for reversible processes} \tag{4.10}$$

In a strict sense, changes in entropy of an open system must not be considered in isolation. Concern should be with total entropy change and this can be obtained by summing the changes of entropy of the system and of the surroundings, i.e.

$$\Delta S_{\text{total}} = \Delta S_{\text{system}} + \Delta S_{\text{surroundings}} \tag{4.11}$$

The example of the oranges in the supermarket illustrated the nature of a spontaneous reaction whereby the ordered pile was disordered very easily. The natural tendency for such an irreversible situation is for entropy to increase. It will be recalled that there is one distribution of oranges which permits the construction of a pyramid—the entropy of such a state is zero. The implication of this example is that irreversible processes cause the increase of total entropy and from equation (4.11) this increase results from changes in the entropy of the system and of the surroundings. It is thus possible for the total entropy to increase whilst the entropy of the system decreases. Thus, in talking about entropy change, the assumption should be that this refers to the system and its surroundings.

Discussion of entropy leads on to a very useful measure called *free energy* which indicates that a process will liberate energy which can be utilized. Such a principle is of course, central to any engine. Besides free energy, there is also unavailable energy, clearly energy which cannot be made available. This unavailable energy is obtained by multiplying entropy (S) by absolute temperature (T). The theoretical justification for this is beyond the scope of this discussion; the product of S and T clearly implies that unavailable energy increases as disorder and temperature increase. Again, with the pile of oranges, maximum unavailable energy occurs when the oranges are scattered over the floor. The total energy is obtained by summing free and unavailable energies, i.e.

$$\text{total energy} = \underset{\text{free energy}}{G} + \underset{\text{unavailable energy}}{TS} \tag{4.12}$$

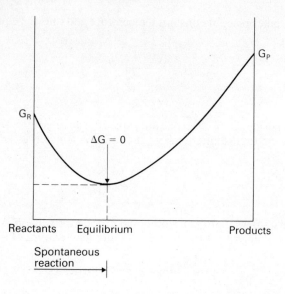

Fig. 4.7 Variation in free energy during a reaction. (from Gymer 1973, p. 591)

Free energy is given the symbol G after Gibbs. The total energy can also be viewed as the sum of the internal energy (E) and the external work performed by the system. Such a measure has already been described since this total energy is the enthalpy (H). Thus equation (4.12) can be written as

$$H = G + TS$$

or

$$G = H - TS \tag{4.13}$$

This states that the free energy is the difference between the total energy and the unavailable energy bound up as heat. The consequences of equation (4.13) are very important throughout science; as entropy values increase, so less energy is available.

Often the prime concern is with the change in free energy (ΔG) rather than its absolute value. Equation (4.13) can thus be modified as follows:

$$\Delta G = \Delta H - T\Delta S \tag{4.14}$$

This assumes constant temperature and pressure conditions.

If two substances are mixed and they immediately react, the free energy decreases to tend to a constant minimum value. At this stage, following the approach of Gymer (1973), there will be no change in free energy and thus ΔG will equal zero. Thus from equation (4.14) the change in enthalpy will be balanced by the change in unavailable energy. These principles are further illustrated in figure 4.7. At the beginning the reactants have a free energy of G_R; as the reaction proceeds, the free energy decreases to a minimum value where the slope on the graph is horizontal and thus ΔG equals zero. At this stage a situation of dynamic equilibrium exists which means that entropy is at a maximum and

$$\text{reactants} \rightleftharpoons \text{products}$$

In this situation the reaction is reversible since it can be made to proceed in either direction by a small change in conditions. The same situation can be achieved by starting with the pure products and their free energy (G_p). The slope of the graph in figure 4.7 is negative between G_R and where ΔG is zero. This part of the graph represents the spontaneous reaction when the reactants are brought together. Thus a negative value of ΔG is characteristic of a spontaneous change whilst ΔG for a system in equilibrium equals zero. When ΔG is positive (between $\Delta G = O$ and G_p in figure 4.7), then a reaction would not take place unless there is an input of energy from the surroundings. Thus the value of ΔG is a useful indicator as to whether a reaction will occur spontaneously. The significance of this in weathering will be demonstrated in chapter 5. ΔG is negative when the free energy of the products is less than the free energy of the reactants; in this situation a chemical reaction will occur. It is useful to note that the value of ΔG, unlike enthalpy change ΔH, varies according to temperature and pressure. This implies that the spontaneity of reactions depends upon temperature and pressure conditions.

The necessary calculations for determining changes in enthalpy from heats of formation have already been introduced. Similar calculations are possible for free energy changes if the standard free energies of formation (ΔG_f°) of reactants and products are known. The superscript $^\circ$ indicates the standard state (gas, liquid or solid) at 25 °C and 1 atm pressure. The free energy of formation of a pure substance is defined as the free energy change which results when one mole of the substance is formed from the elements at 1 atm pressure and at 25 °C. Values for ΔG_f° can be obtained from the *Handbook of Chemistry and Physics* (1974) edited by Weast, but some values are given in table 4.1. A worked example modified from Masterton and Slowinski (1973) may aid understanding. The problem is to determine the free energy change ΔG° for the following reactions:

(a) $\qquad Al_2O_3 + 3H_2(g) \rightarrow 2Al(s) + 3H_2O(l)$

(b) $\qquad 4NH_3(g) + 5O_2(g) \rightarrow 4NO(g) + 6H_2O(l)$

The general equation is

$$\Delta G^\circ = \sum \Delta G_{f\,(products)}^\circ - \sum \Delta G_{f\,(reactants)}^\circ$$

It is also important to realize that the free energies of formation of elementary substances are by definition, zero. Values of ΔG_f° for the compounds are given in table 4.1.

(a) $\qquad \Delta G^\circ = 2\Delta G_f^\circ Al(s) + 3\Delta G_f^\circ H_2O(l)$

$\qquad\qquad - [\Delta G_f^\circ Al_2O_3(s) + 3\Delta G_f^\circ H_2(g)]$

$\qquad\qquad = 3(-237 \cdot 19) - (-1576 \cdot 41) \text{ kJ mol}^{-1}$

$\qquad\qquad = 864 \cdot 84 \text{ kJ mol}^{-1}$

This reaction does not occur at 25 °C because $\Delta G^\circ > 0$.

(b) $\qquad \Delta G^\circ = 4\Delta G_f^\circ NO(g) + 6\Delta G_f^\circ H_2O(l) - 4\Delta G_f^\circ NH_3(g)$

$\qquad\qquad = 4(86 \cdot 69) + 6(-237 \cdot 19) - 4(-16 \cdot 64) \text{ kJ mol}^{-1}$

$\qquad\qquad = -1009 \cdot 82 \text{ kJ mol}^{-1}$

Thus this reaction is spontaneous at 25 °C and 1 atm.

It is now possible to summarize the concepts associated with the second law of thermodynamics. The first law is concerned with the conservation of energy whilst the second is concerned with its degradation. Various specifications of the second law can be found in the literature and some are reproduced below. The advantage of quoting these statements is not only to describe the law, but also to summarize some of the principles which have emerged in this chapter.

Second law of thermodynamics

'For spontaneous change to occur in an isolated system, the entropy must increase, that is, ΔS must be greater than zero.' (Sienko and Plane 1974, pp. 372–3).

'No process involving an energy transformation will spontaneously occur unless there is a degradation of the energy from a concentrated form into a dispersed form.' (E. P. Odum 1971, p.37).

'Because some energy is always dispersed into unavailable heat energy, no spontaneous transformation of energy (light, for example) into potential energy (protoplasm, for example) is 100 per cent efficient.' (E. P. Odum 1971, p. 37).

'The free energy change of any process that occurs spontaneously is negative.' (Anderson, Ford and Kennedy 1973, p. 194).

'Systems tend to move toward equilibrium.' (Anderson, Ford and Kennedy 1973, p. 194).

An appropriate way to conclude this chapter is to demonstrate some of these principles with reference to specific examples drawn from the physical environment—ecosystems, soil moisture and rivers. Details of thermochemistry are exemplified in chapter 5 when weathering is discussed.

4.3 Application of energy principles to the physical environment: three examples

(1) Ecosystems

An energy approach to ecosystem analysis has become very common, due primarily to the pioneering work of E. P. and H. T. Odum, indeed the latter has also applied such principles to society as a whole (H. T. Odum 1971). Morris (1973, p. 10) quoting R. B. Miller defines ecosystems as 'open systems, comprising plants, animals, organic residues, atmospheric gases, water and minerals which are involved together in the flow of energy and the circulation of matter.' The analysis of movement of matter within ecosystems can be tackled by identifying the various constituent *cycles*; the best known perhaps are the nutrient cycles of phosphorous, nitrogen, sulphur, etc. But the dynamics of ecosystem evolution and eventual equilibrium can only be understood when the energetics of the system are analysed. In contrast to nutrients, energy enters an ecosystem, is transferred through it and then is lost to the system. The major input of energy to all ecosystems is solar radiation which is utilized by primary consumers to convert the energy to food by photosynthesis. However, only a small proportion of solar radiation is converted into usable form by photosynthetic processes. It is possible to compare the rate of receipt of solar radiation at the ground surface with gross primary productivity which is the total rate of photosynthesis, including the organic matter used up in respiration (E. P. Odum 1971). Odum gives figures to compare solar radiation (11 088 kJ m^{-2} d^{-1}) and gross primary productivity for sugar beet in England (845 kJ m^{-2} d^{-1}). The magnitude of the difference between these figures indicates the inefficiency of photosynthesis in converting energy. The food or energy produced by photosynthesis can be passed through the ecosystem by a food chain or if the situation is complex, by a food web. Each step in the transfer of energy is called a trophic level and a simple example of a chain is primary producers (autotrophs), herbivores and carnivores. The principles governing such energy transfer as well as the efficiency of energy exchange are of particular importance both in a theoretical and practical sense. If greater output is required from an ecosystem, then energy has to be added to the system. The usual method of improving agricultural production is by applying energy subsidies in the form of fertilizers.

An excellent example of an ecosystem is a pond and figure 4.8 illustrates the principal food chains of a particular pond in Georgia which is managed for sports fishing (E. P. Odum 1971). For simplification the successive ingested quantities of energy are indicated for the various trophic levels. At the surface 30.5432×10^5 kJ m^{-2} a^{-1} of available solar radiation are received and 30.962×10^3 kJ m^{-2} a^{-1} of these are utilized by phytoplankton, which are fed upon by zooplankton and bloodworms. Predaceous diptera larvae eat the zooplankton whilst sunfish have a more varied diet eating the larvae,

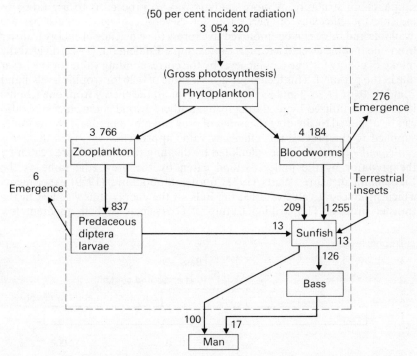

Fig. 4.8 Model of the principal food chains in a Georgia pond managed for sports fishing. Estimated energy inputs are in kilojoules per square metre per year. (from Odum E. P. 1971, p. 70).

zooplankton and the bloodworms. The fishermen's objective is the bass which eat the sunfish. A certain amount of energy exchange takes place across the boundaries of the system as indicated in the figure. In terms of maximizing catch from the pond, it is clear that bass should be eliminated and fishermen just fish for sunfish; this demonstrates the fundamental principle that the optimum management of ecosystems involves maintaining only the minimum number of trophic levels, a practical consequence of the second law of thermodynamics. However, in this context sport is only possible if the catch is reasonably sized bass rather than small sunfish. Another possibility appears to be to remove the predaceous larvae because of their tiny contribution of energy to the sunfish, but such action might disturb the stability of the ecosystem.

Various measures of energy exchange have been developed in order to express the transfer of energy through the various trophic levels, these techniques are summarized by Odum (1971) and Watts (1971). One graphic method is to construct a *pyramid of energy* which is akin to a population pyramid. The base is drawn in proportion to the amount of energy injested by the primary producers in the ecosystem and the next tier corresponds to the amount of energy which the primary consumers ingest. Figure 4.8 shows the result from the Georgia pond example; the zooplankton and bloodworms are grouped to constitute the second trophic level whilst the third comprises the larvae and the sunfish. In a strict sense these latter two components should not be grouped since the sunfish eat the larvae. Thus difficulties arise in the construction of a pyramid of energy when the ecosystem does not have a simple chain structure. A more analytic way to proceed is to use ratios to indicate the efficiency of energy transfer. Several ecological efficiencies are available and these can be employed to express the efficiency of energy transfer from one trophic level to another. For example Lindemann (1942) divided the intake of energy for one trophic level by the corresponding value for the lower one in the pyramid. This technique gives a value of 0·26 for trophic levels 1 and 2 and 0·29 for levels 2 and 3 of figure 4.9. Not all the energy from one trophic level may be utilized by predators from another. A food chain efficiency ratio can be obtained by dividing the joules of prey eaten by a predator by the joules supplied to the prey. If this efficiency value approaches one, then the *gross ecological efficiency* can be calculated by dividing the joules of prey eaten by the predator by the joules of food eaten by the prey—the same as the Lindemann measure. Watts (1971) quotes Slobodkin's (1959) experiments which produced a value of 0·13; the bulk of the energy thus was lost in the transfer process. The striking feature of ecosystems is their apparent low

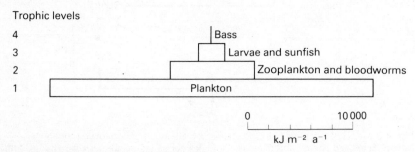

Fig. 4.9 Ingested kilojoules per trophic level.

efficiencies in contrast to mechanical systems. E. P. Odum (1971) points out that such comparisons are not valid strictly speaking since ecosystems are self-repairing; service and repair of cars ought to be included in calculating the efficiency of car engines. Indeed E. P. Odum (1971) and H. T. Odum (1971) develop the argument that low efficiency is necessary for maximum output and such a suggestion based upon an energy analysis has marked implications for the management of ecosystems.

(2) Soil moisture

The second example to demonstrate the applicability of energy principles is far more specific—soil moisture—yet an understanding of the principles governing soil moisture is essential for any analysis of soil–plant relationships, of soil temperature or of soil erosion. The modern approach to soil moisture is to appreciate that the retention and movement of water, its uptake by plants, and its ultimate loss to the atmosphere are all energy related phenomena (Brady 1974).

The notion of energy and soil water requires some elaboration. The idea that water in a beaker contains a finite quantity of energy has already been introduced and thus talk of soil water energy should come as no surprise. The energy of soil water is a reflection of the forces which are influencing the water; if there is a resultant force vector, then the water will move from areas of higher specific energy (energy per unit quantity) to areas where it is lower. A measure of this specific energy is difficult to obtain and the task is easier if the difference in energy is determined between soil water and a hypothetical reservoir of pure free water at a specific elevation. This difference is termed the *total potential of soil water* which is defined as the amount of work which must be done per unit quantity of pure water to transfer from the reservoir a quantity of pure water to the relevant point in the soil. Such a process must be carried out at constant temperature and in a reversible manner. The total potential of water in unsaturated soil always has a negative value if the total potential of water in the reference reservoir is set at zero.

These principles may seem to be very theoretical, but as has been suggested, it is spatial and temporal variations in total soil water potential which are so very relevant to a wide variety of problems. The term total potential implies that it results from various constituent potentials which can be more readily appreciated. In fact the total potential of soil water (ϕ_t) can be explained primarily by three types of potential; gravitational potential (ϕ_g), matric potential (ϕ_m), and osmotic potential (ϕ_o) (Hillel 1971), i.e.

$$\phi_t = \phi_g + \phi_m + \phi_o \qquad (4.15)$$

These require to be explained in turn.

Gravitational potential

This type of potential is the easiest to understand because it is potential energy. In principle, gravitational potential is the amount of work per unit quantity of pure water which has to be expended in order to transport water from the reference reservoir to the appropriate elevation in the soil (International Society of Soil Science 1963):

$$\phi_g = \rho_w\, g\, h$$

where ϕ_g is gravitational potential per unit volume, ρ_w is the density of water, g is the acceleration due to gravity and h is the height of the point in question above the reference reservoir. Clearly gravitational potential causes soil moisture to move downwards; such movement is often constrained by the other two main types of potential. This can also be expressed by noting that the effect of height above the reference level is to increase the free energy of soil water, but matric and osmotic potentials reduce this free energy.

Matric potential

The formal definition of this potential is based on the energy required to transport a *soil solution* from the reference level to a point in the soil at the same elevation. These special conditions are required for the definition in order to exclude gravitational and osmotic effects. This potential results from the nature of the mineral matrix which is able to exert specific retentive forces on soil moisture. In particular the solid particles and the intervening pores are able to retain water by absorption and capillarity. In order to explain these brief mention must first be made of surface tension.

The interface between a liquid and a gas can be viewed as a taut skin; in the case of free water there are forces between the water molecules, but at the surface there are forces between molecules of water and atmospheric molecules. These surface forces are less than those within the water, and so there is a net tension over the surface. Surface tension is the force per unit length which must be applied to overcome molecular forces. In practice it is determined by measuring the force necessary to pull a platinum ring off the surface of a liquid. The magnitude of this surface tension depends primarily upon the type of solution and its temperature. Surface tension is closely connected with capillarity which can be demonstrated by considering the partial insertion of a thin open glass tube into a container of water. On insertion water enters the tube and adhesion between water molecules and the glass causes a slight rise of the liquid round the inner side of the tube to form a meniscus. Because the surface of the liquid in the tube is not horizontal, surface tension forces have an upward force component. This results in the rise of water in the tube until the force arising from the uplifted mass of water equals the resultant force from adhesion and surface tension. The height of capillary rise in the tube can be approximately found by the equation:

$$h \simeq \frac{2\gamma}{r\rho g} \qquad (4.16)$$

where h is the height of capillary rise, γ is the surface tension, r is the radius of the meniscus, ρ is the density of liquid and g is the acceleration due to gravity. The relevance of this to soils is that water can rise by capillarity depending upon the size of the interconnecting pores. In practice the height predicted from the above equation is not achieved because a variety of other factors must be taken into account.

Water molecules, as well as being attracted to solid surfaces (adhesion) are also attracted to each other by hydrogen bonding (chapter 7), in a process known as *cohesion*. The combination of adhesive and cohesive forces thus causes adsorption. In this context adsorption is taken to mean only the physical and reversible attraction of water to solid particles. Figure 4.10

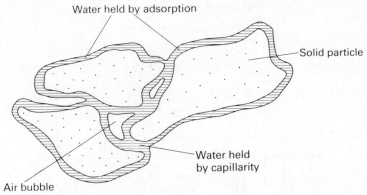

Water held by adsorption

Solid particle

Water held
by capillarity

Air bubble

Fig. 4.10 Soil water held by capillarity and adsorption. (after Brady 1974, p. 171)

illustrates how water is held in the soil by capillarity and adsorption. The combined effect of adsorption and capillarity is the matric potential.

Osmotic potential

This is defined in terms of the amount of work required to transport pure water from a reservoir to another at the same elevation containing a soil solution. The International Society of Soil Science (1963) again gives a formal definition, but the principles of osmosis are easier to understand by consideration of the U-tube situation as in figure 4.5. The reasons for the difference in elevation of the levels on either side of the membrane have been discussed and this difference corresponds to the osmotic potential.

Brady (1974) has produced a useful diagram which neatly illustrates the relationships between matric, osmotic and total soil water potentials (figure 4.11). For simplification the gravitational potential is not included. The moist soil in the centre is separated from pure water on the right by a semipermeable membrane. The combined effect of the matric and osmotic potentials causes the elevation of the mercury which represents the soil water potential. To the left of the wet soil the design is such as to separate the osmotic and matric potentials by having a column of soil solution wedged in between two semipermeable membranes. It is clear from the mercury levels in this case that the total soil water potential equals the sum of the matric and osmotic potentials.

Total soil water potential in a strict sense has the dimensions of energy, but it is common to express such a potential in terms of pressure or suction. A useful measure of pressure is the length of a column of water required to produce the necessary suction, for instance, a column of water 1033 cm long produces a pressure of $10 \cdot 13$ N cm^{-2} and is equivalent to a length of 76 cm of mercury or 1 atm. It has been found useful to employ a logarithmic scale to represent suction since the lengths of columns of water soon reach very large values. The pF scale which is the logarithm of the length of tube in centimetres is adopted. Thus a soil which is almost saturated and is under slight suction, has a negative pressure of 10 cm of water or $1 \cdot 0$ pF. A soil with partial evacuation of its pores must be under a certain suction, say 1000 cm or $3 \cdot 0$ pF. The amount of water in a soil varies according to the applied suction and pore size distribution and the

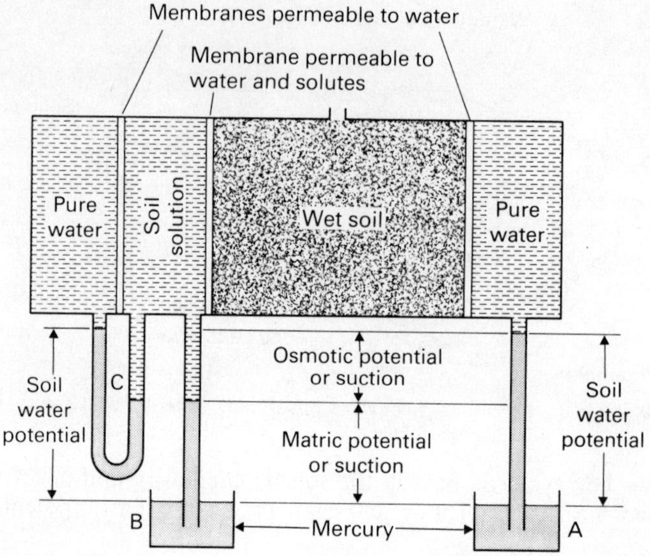

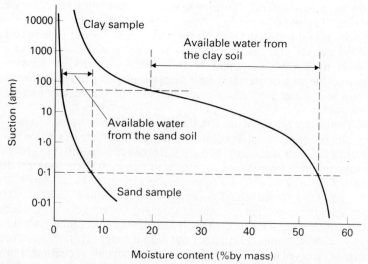

Membranes permeable to water

Membrane permeable to water and solutes

Pure water

Soil solution

Wet soil

Pure water

Soil water potential

C

Osmotic potential or suction

Matric potential or suction

Soil water potential

B

Mercury

A

Fig. 4.11 Relationship among osmotic, matric, and the combined soil water potentials. The system is assumed to be at equilibrium and at constant temperature. The combined potential from both osmotic and matric attractive forces (soil water potential) is evidenced by the attraction of pure water (right) through the membrane and is measured at equilibrium by the rise in mercury from vessel A. The osmotic potential is given by the difference in pressure between the pure water (left) and the soil solution (measured by manometer C). The matric potential is the difference between the combined and osmotic potentials and is measured by the height of rise of the mercury in vessel B. Since both osmotic and matric potentials are negative, they are sometimes referred to as suctions or tensions. The gravity potential is not shown in this diagram. (from Brady 1974, p. 170)

Suction (atm)

10000

1000

100

10

1·0

0·1

0·01

Clay sample

Available water from the clay soil

Available water from the sand soil

Sand sample

0   10   20   30   40   50   60

Moisture content (%by mass)

Fig. 4.12 Soil moisture retention curve for a sand and clay. The suction range for available water to plants is shown.

relationship between moisture content and suction can be expressed as a soil moisture retention curve. Figure 4.12 illustrates such curves for two contrasting soils, a sand and a clay: with the former the large pores drain under the slight application of suction whilst a clay can retain much moisture even at high suctions. The amount of water available to plants can be obtained from such curves by knowing that plants can exert suctions up to the order of 15 atm or 4·2 pF.

So the amount of moisture in soil can be approached by considering the constituent potentials as outlined above. The manner in which soil yields its moisture depends on the applied suction and pore size distribution which are expressed in the soil moisture retention curve.

(3) Rivers

The two laws of thermodynamics have been applied by Leopold and Langbein (1962) to the evolution of rivers and thus to landscape development. They analyse the energetics of a river system by treating it as a series of perfect engines which function between heat sources and sinks; each engine yields a quantity of work. Equations are obtained to express the change of temperature (analogous to height) with horizontal distance in terms of entropy change and thus the longitudinal profile of a river is predicted when entropy is maximized. Instead of the detailed quantitative argument of Leopold and Langbein (1962), a qualitative approach as given by Leopold (1962) in another publication now follows since this illustrates in general terms the applicability of thermodynamic principles to rivers.

The relevance of energy principles to rivers was indicated at the end of chapter 3. A small isolated volume of water as it moves from the source of a river to the mouth decreases in height and thus drops in potential energy. The *total* energy of this small volume of water, if it can be considered in isolation, must remain constant according to the first law of thermodynamics; however as the water drops in height, so the *available* energy decreases, a consequence of the second law. A decrease in available energy implies an increase in entropy. Thus rivers, or indeed any slopes, as they decrease in elevation away from their sources or summits, increase in entropy. The nature of this increase ultimately influences the form of these slopes. When constancy of form is achieved (equilibrium), entropy is maximized.

Rivers, like all geomorphic systems, are open in the sense that energy or matter can be added or removed throughout their lengths. Friction along the channel can lead to the slight generation of heat which is rapidly lost along the channel margins. Precipitation is another input as well as heat from solar radiation. Losses arise, for example, from evaporation and conduction. Nevertheless, it is conceivable for all these inputs and outputs to the system as well as the system itself to achieve equilibrium—in particular such a system displays dynamic equilibrium. In this situation entropy is constant and maximized to mean that entropy change is zero. This applies to the system and its surroundings (see equation (4.11)) meaning that the rate of outflow of entropy from the river to its surroundings (represented by dissipation of heat) is balanced by the increase of entropy within the river. The statistical nature of entropy has been described and was shown to be dependent on the probability of the occurrence of particular energy distributions. In the case of rivers, the change in energy depends on the change in elevation. Thus the increase in entropy within the river can be equated with the change in elevation of the river—the energy gradient. Leopold (1962) concluded that the most probable sequence of energy losses in unit lengths of stream channel correspond to a uniform increase in entropy. In such a situation assuming no constraint on stream length, the long profile of a river tends to an exponential form. This can be compared with the form of actual rivers. For example, Jones (1924) found that the best function to fit the long profile of rivers in south Wales was

$$H = c - k \log (x+a) + b(x+a)$$

where H is height, x is horizontal distance and a, b and c are constants.

The discussion of the analysis of Leopold and Langbein is a simplified form of their original paper. Yang (1972) has developed their approach and has proposed two basic laws in fluvial geomorphology using the analogy of entropy in thermodynamics. 'The (first) law of average stream fall states that under the dynamic equilibrium condition the ratio of the average fall between any two different order streams in the same river basin is unity. The (second) law of least rate energy expenditure states that during the evolution towards its equilibrium condition, a natural stream chooses its course of flow in such a manner that the rate of potential energy expenditure per unit mass of water along this course is a minimum.' (Yang 1972, pp. 321–2). The behaviour of a drainage network can be explained in part by these laws. As well as seeking to minimize the expenditure of energy, rivers also adjust so as to minimize discontinuities in energy loss along their lengths. The alteration of pools and riffles in a straight channel introduces such discontinuities and the response of rivers is to develop meanders in order to increase the loss of energy in pool sections; thus the explanation of meanders as given by Leopold, Wolman and Miller (1964) is in terms of a river seeking a smooth energy grade line.

5
Chemical reactions and equilibrium

Chemical changes in the physical environment result from chemical reactions. In chapter 2 the nature of chemical equations was outlined and in the preceding chapter the basic concepts of thermochemistry were introduced. A chemical reaction is possible if the free energy of the products is less than the free energy of the reactants. The magnitude of this difference (ΔG) depends upon the nature of the reactants and products as well as temperature and pressure conditions. These factors thus influence the rate of reactions, but equally important are the concentrations of reactants as well as the presence of any substances which assist but do not participate in the reaction. These substances are called *catalysts* if they are inorganic or *enzymes* if organic. The rates of chemical reactions are akin to geomorphological processes in that they can vary from being almost instantaneous to being measurable only on the geological time scale. Another important point to note is that chemical reactions though perhaps represented in one equation, involve a series of steps.

5.1 Introductory principles associated with reactions

Homogeneous reactions take place between substances in the same phase; with this type the rate depends on the nature and concentration of the reactants. More usually in the physical environment reactions take place between substances in different phases—these are called heterogeneous reactions. With this type the rate of change is influenced by the area of contact between the phases. The rate of solution of silt-sized fragments of limestone is clearly going to be greater than the rate for sand-sized fragments and this is the reason why farmers apply ground limestone composed of particles of various sizes to their fields in order to make immediate the effect as well as to prolong the benefits of the limestone. Another way to approach reactions is to consider them according to the number of reactants; first-order reactions are the simplest, whereby one substance spontaneously changes into another. The radioactive decay of $^{14}_{6}C$ is an example. A first-order reaction can be represented as

$$A \rightarrow B \tag{5.1}$$

The rate of the process can be determined by measuring how much of B is produced in an interval of time (Δt), i.e.

$$\text{rate} = \frac{\Delta B}{\Delta t}$$

The increase in substance B is paralleled by a decrease in A and thus the rate can also be expressed as

$$\text{rate} = -\frac{\Delta A}{\Delta t}$$

The minus sign is used to indicate a decrease in substance A. The rate at which molecules of substance B are being formed will clearly depend upon the number of molecules of substance A. This result is more formally expressed in the *law of mass action* which states that the rate of a reaction is directly proportional to the concentration of the reagents (Stamper and Stamper 1971). A measure of concentration is thus necessary—an amount of substance in a given volume. The amount is usually expressed in moles whilst a litre is the commonly used unit of volume. Thus one mole of a substance dissolved in one litre of water gives a solution of $1 \cdot 0 \ \text{mol} \ l^{-1}$. The convention is to indicate the concentration of a substance by the use of square brackets. Hence $[A]$ represents the concentration of A in moles per litre. For the first-order reaction expressed in equation (5.1), the rate of change of A into B is proportional to $[A]$, i.e.

$$\text{rate} \propto [A]$$

or

$$\text{rate} = k_1[A] \tag{5.2}$$

where k_1 is a constant known as the specific rate constant. Each first-order reaction has a characteristic value of k_1 which is also dependent upon temperature and the presence of catalysts.

Equilibrium occurs in a chemical reaction when the forward reaction from reagents to products is exactly balanced by the backward reaction of products to reactants. In the simple case presented in equation (5.1), equilibrium can be expressed as

$$A \rightleftharpoons B$$

It is thus possible to express this rate of backward reaction as

$$\text{rate} = k_2[B] \tag{5.3}$$

where k_2 is the specific rate constant.

A useful ratio is obtained when the two rate constants are divided, i.e.

$$K = \frac{k_1}{k_2} \tag{5.4}$$

where K is called the equilibrium constant, k_1 is the forward rate constant and k_2 is the backward rate constant. The value of K indicates the extent to which products are favoured over reactants or K is the ratio of the tendency of the products to be formed to the tendency of the reactants to be formed (Anderson, Ford and Kennedy 1973). The value of K for any reaction, like rate constants, depends upon temperature. In figure 4.7 a state of equilibrium is illustrated when the change in free energy (ΔG) is zero. The rate of spontaneous reaction towards equilibrium is expressed in the rate of change of free energy (the angle of slope between G_R and $\Delta G = 0$ in figure 4.7). From the description of the equilibrium constant, a relationship between K and free energy change is to be expected.

If equation (5.2) is divided by equation (5.3), the following is obtained:

$$\frac{\text{rate}}{\text{rate}} = \frac{k_1}{k_2}\frac{[A]}{[B]}$$

At equilibrium, the rates are equal and thus

$$K = \frac{k_1}{k_2} = \frac{[B]}{[A]} \tag{5.5}$$

This illustrates the rule that equilibrium constants are always presented so that the concentration of products on the right-hand side of a chemical equation is the numerator whilst the concentration of reactants is the denominator.

The calculation of rates becomes a little more complicated when second-and higher-order reactions occur; these refer to reactions in which two or more reactants are involved respectively. With first-order reactions an increase in concentration in reactant leads to a linear increase in rate; this is not so with higher-order reactions. This can be understood by recalling that the molecules in liquids and gases are constantly moving at high speeds and are nearly always in collision—reactions result from such collisions. With first-order reactions the probability of collision between molecules of one type depends in linear fashion on their numbers. When two or more types of molecules are present, the chance of two different molecules being at the same point at the same time is equal to the product of the constituent probabilities. Suppose a second-order reaction can be expressed as follows:

$$aA + bB \rightleftharpoons cC + dD \tag{5.6}$$

where A, B, C and D are the reactants and products, and a, b, c and d are the respective numbers of moles of these. The rate per unit volume of the forward reaction in this case is thus a power function, viz.

$$\text{rate} = k_1[A]^a[B]^b \tag{5.7}$$

This equation is a more general form of the rate law. If there are more than two reactants, then the concentration of these to the appropriate power is tacked on to the right-hand side of the equation, i.e.

$$\text{rate} = k_1[A]^a[B]^b[C]^c[D]^d \tag{5.8}$$

The same approach applies to the backward reaction and thus the equilibrium constant for equation (5.6) is obtained as follows:

$$K = \frac{[C]^c[D]^d}{[A]^a[B]^b}$$

Of particular concern to the physical geographer is the effect of climate on reaction rates and the two critical climatic characteristics are temperature and moisture. The essential *presence* of water for practically all chemical processes in the physical environment will be demonstrated later in this chapter; in contrast temperature has a marked influence on the *rate* of reactions. The value of K, the specific rate constant, depends not only on the nature of the reaction, but also upon the temperature. Experimentation is necessary to determine K, but a very rough rule which is sometimes given is that with every $10 C°$ rise in temperature, the reaction is two or three times faster (Sienko and Plane 1976).

This effect of temperature can be understood by recalling that the rate of movement of molecules or ions is dependent upon temperature; the collision rate thus increases with higher temperatures. However, it is important to stress that all collisions do not result in change—many collisions must take place without changes in electron or molecular configuration otherwise all reactions would be over in an instant. Ions or molecules can bump into one another without causing change, but if these collisions are more severe, then changes will result. Sometimes an energy input is applied (energy of activation) so that such changes are encouraged. Many substances on mixing will react because there is sufficient internal energy for changes to result from ionic or molecular collisions, but with others, an extra input is necessary to make the reaction possible. Clearly the magnitude of this supplementary energy input depends on the particular reaction. The example of photosynthesis was quoted in chapter 4 with the energy subsidy being provided by solar radiation. Photosynthesis, being a very slow reaction, requires a major input of energy of activation. In contrast, fast reactions require little such support.

A common misconception is that in a chemical reaction, ultimately all reactants are utilized to yield products; this is not the case. Instead, in any reaction, there is a tendency towards a chemical equilibrium which is characterized by constancy in quantity both of reactants and products. Consider the simple case of reactants A and B combining to produce C. If the amounts of A and B were measured as the reaction proceeded, then they would decline through time, but would tend towards common individual values (figure 5.1). After time t_e, a state of chemical equilibrium exists. This does not imply that the reaction has terminated; instead the production of C by A and B is being balanced by the production of A and B by the split up of C. So at equilibrium the reversible process

$$A + B \rightleftharpoons C$$

is balanced by the forward and reverse reactions. States of chemical equilibrium, in the same way as physical dynamic equilibrium, are characterized by specific constant values for the factors involved. Take one atmospheric component, carbon dioxide, CO_2; the quantity of CO_2 should be constant if the processes which produce and utilize CO_2 are in equilibrium. The significance of the equilibrium constant can be further demonstrated by considering water which is composed of a mixture of water molecules (H_2O), hydrogen cations (H^+) and hydroxyl anions (OH^-). At equilibrium the reversible process can be expressed as

$$H_2O \rightleftharpoons H^+ + OH^{-1} \tag{5.9}$$

In practice hydronium ions (H_3O^+) exist rather than hydrogen ions, but it is customary to treat H_3O^+ ions as though they are H^+ ions. Equation (5.9) has the implication that pure water when in equilibrium has characteristic

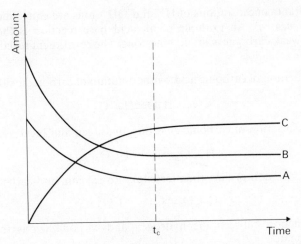

Fig. 5.1 The reaction of substances A and B to produce C. After time t_c, chemical equilibrium exists.

concentrations of H^+ and OH^- ions and H_2O molecules. The *ion product* (K_w) is obtained by multiplying the concentrations of the ions, i.e.

$$K_w = [H^+][OH^-]$$

At 25 °C the value of K_w is $1 \cdot 00 \times 10^{-14}$ mol^2 l^{-2}. The concentrations of H^+ and OH^- ions must be the same, given equation (5.9), and these must thus have a numerical value of half of K_w. In other words, with pure water, H^+ and OH^- ions both have concentrations of $1 \cdot 00 \times 10^{-7}$ mol l^{-1}. This has led to the definition of the pH scale whereby the pH is stated to be the negative of the logarithm of the concentration of H^+ ions, i.e.

$$pH = -\log[H^+] \tag{5.10}$$

Remembering that $\log_{10} 100 = 2$, then the pH of pure water is 7 which is defined as the neutral value. High concentrations of H^+ ions result in pH values less than 7—acid conditions. This is understood by remembering that a number such as 10^{-6} is bigger than 10^{-7}. Similarly alkaline conditions result when pH values are over 7. Because the pH scale is logarithmic, a pH of 5, for example, is twice as acidic as a pH of 6.

5.2 Acids and bases

It has been recognized since the early days of chemistry that certain substances called acids react with metals to produce hydrogen. Such a reaction is because of the excess concentration of H^+ ions. Substances which when dissolved in pure water give a higher concentration of H^+ ions are called *acids*. Similarly substances which when dissolved in pure water give a higher concentration of OH^- ions are called *bases*. Acids and bases react to produce *salts*, for example hydrochloric acid (HCl) and sodium hydroxide (NaOH) react to produce sodium chloride (NaCl), viz.

$$HCl + NaOH \rightarrow NaCl + H_2O \tag{5.11}$$

If the resultant concentrations of H^+ and OH^- ions are equal, then the salt solution is *neutral*. An example of an acid–base reaction is the effect of rainwater (weak carbonic acid) on limestone. The constituent reactions are as follows:

(1) The formation of carbonic acid by the solution of carbon dioxide in water

$$CO_2 + H_2O \rightleftharpoons H_2CO_3 \qquad (5.12)$$

(2) The dissociation of carbonic acid into hydrogen and bicarbonate ions

$$H_2CO_3 \rightleftharpoons H^+ + HCO_3^- \qquad (5.13)$$

(3) The dissociation of limestone (calcite) into calcium and carbonate ions

$$CaCO_3 \rightleftharpoons Ca^{2+} + CO_3^{2-} \qquad (5.14)$$

(4) The reaction between the hydrogen and carbonate ions to produce bicarbonate

$$H^+ + CO_3^{2-} \rightleftharpoons HCO_3^- \qquad (5.15)$$

The left- and right-hand sides of equations (5.12)–(5.15) can be added to give the following equation after cancellation of common terms:

$$CO_2 + H_2O + CaCO_3 \rightleftharpoons Ca^{2+} + 2HCO_3^- \qquad (5.16)$$

This can also be expressed as:

$$H_2CO_3 + CaCO_3 \rightleftharpoons Ca(HCO_3)_2 \qquad (5.17)$$

The calcium and bicarbonate ions are lost in solution; if there is evaporation, the reverse process operates and calcium carbonate is precipitated, for example stalactites and stalagmites. This acid–base reaction illustrates that chemical reactions are often composed of a series of reactions or equilibria.

A limitation of the definitions as given of acids and bases is that an aqueous solution is necessary for dissociation into ions containing hydrogen or hydroxide ions. A more general approach is to define an acid as a proton donor whilst a base is a proton acceptor. Of course a proton is a hydrogen ion so that an acid is still a substance which yields a hydrogen ion, but not necessarily in solution.

Acids and bases vary in their strength, a reflection of their differing ability to yield or accept protons. If consideration of strength is limited to the aqueous situation, then the significant quality is the ability to dissociate. Acids and bases are strong if they dissociate very well into their constituent ions; with weak acids or bases, such dissociation is only partial. The strong hydrofluoric acid (HF) dissociates into hydrogen and fluorine ions, viz.

$$HF \rightleftharpoons H^+ + F^- \qquad (5.18)$$

The equilibrium constant (K) for this reaction is

$$K = \frac{[H^+][F^-]}{[HF]}$$

if the role of water is omitted. In particular K is given the symbol K_a since it is called the acid ionization constant of hydrofluoric acid. At 25 °C the value of

K_a for hydrofluoric acid is $6 \cdot 9 \times 10^{-4}$ mol l^{-1}. Similarly the acid ionization constant for carbonic acid (see equation (5.13)) can be obtained from

$$K_a = \frac{[H^+][HCO_3^-]}{[H_2CO_3]}$$

In this case K_a has a value of $4 \cdot 2 \times 10^{-7}$ mol l^{-1} at 25 °C. For simplicity the second ionization step associated with carbonic acid (the bicarbonate ionizing into hydrogen and carbonate ions) is ignored. The strength of bases is expressed in a similar way, but with K_b representing the *base ionization constant*. For example, the base, ammonium, reacts with water as follows:

$$NH_3 + H_2O \rightleftharpoons NH_4^+ + OH^- \qquad (5.19)$$

If water is excluded, then K_b for ammonia is determined from

$$K_b = \frac{[NH_4^+][OH^-]}{[NH_3]}$$

In this case K_b has a value of $1 \cdot 8 \times 10^{-5}$ mol l^{-1} at 25 °C. Such a small value indicates that there is very little ammonium (NH_4) in a solution of ammonium hydroxide; nearly all is in the form of ammonia (NH_3). The values of K_a and K_b indicate the strength of acids and bases; the smaller the value, the weaker the acid or base. Values for selected acids and bases are given in table 5.1. A value of around $1 \cdot 0$ for K_a specifies a moderately strong acid whilst values of about $10 \cdot 0$ and over distinguish very strong acids (for example, sulphuric acid, H_2SO_4). Care should be taken to distinguish between the strengths of acids or bases and their concentrations; the former express the extent to which ionization can take place at equilibrium whilst the latter refer to the amount of the substance present in a unit volume of water.

In practice it is often desirable to know whether a solution is acidic or alkaline and this can be determined by using an *indicator* which has the

Table 5.1 Ionization constants at 25°C for weak acids and weak bases in aqueous solution (data for K_a from Gymer 1973, p. 300 and data for K_b from Weast 1974, p. D-128).

Acids	Formula	Ionization constant (K_a) (mol l^{-1})
acetic acid	$CH_3COOH(HAc)$	$1 \cdot 8 \times 10^{-5}$
carbonic acid	H_2CO_3	$4 \cdot 2 \times 10^{-7}$
	HCO_3^-	$4 \cdot 8 \times 10^{-11}$
hydrofluoric acid	HF	$6 \cdot 9 \times 10^{-4}$
sulphuric acid	H_2SO_4	~ 10
	HSO_4^-	$1 \cdot 2 \times 10^{-2}$
sulphurous acid	H_2SO_3	$1 \cdot 3 \times 10^{-2}$
	HSO_3^-	$5 \cdot 6 \times 10^{-8}$

Bases	Formula	Ionization constant (K_b) (mol l^{-1})
ammonium hydroxide	NH_4OH	$1 \cdot 79 \times 10^{-5}$
calcium hydroxide	$Ca(OH)_2$	$3 \cdot 74 \times 10^{-3}$
lead hydroxide	$Pb(OH)_2$	$9 \cdot 6 \ \times 10^{-4}$
silver hydroxide	$AgOH$	$1 \cdot 1 \ \times 10^{-4}$
zinc hydroxide	$Zn(OH)_2$	$9 \cdot 6 \ \times 10^{-4}$

property of assuming a colour characteristic of the conditions. For example, the well known indicator *litmus* is red in an acid solution and blue in an alkaline one. Other indicators can be selected which change colour at various pH values; for example, methyl orange is red below 4·0 pH whilst it is yellow above. With certain indicators it is possible to estimate the value of pH within defined ranges; as an example the use of a barium sulphate soil test kit permits the determination of pH to within 0·5. For greater precision, a pH meter has to be used.

Often a striking characteristic of a system is its pH stability, achieved by *buffering* processes. These processes mean that any change in pH resultant upon the addition of H^+ or OH^- ions is minimized. A buffer solution, for example, can be made through mixing a weak acid solution with a salt solution containing the same anion as the acid. Gymer (1973) considers the natural processes which regulate the pH value of sea water to illustrate the nature of buffering processes. The pH value of sea water is remarkably constant at a value of 8·1 though small variations are possible. Bicarbonate (HCO_3^-) and carbonate (CO_3^{2-}) occur in abundant quantities in sea water. These anions in addition to carbonic acid (H_2CO_3) are the essential components of the buffering processes. The key reactions are presented in equations (5.12)–(5.15). The addition of an acid is tackled by the reactions

$$H^+ + CO_3^{2-} \rightleftharpoons HCO_3^-$$

and

$$H^+ + HCO_3^- \rightleftharpoons H_2CO_3 \rightarrow H_2O + CO_2$$

Excess H^+ can also be tackled if solid $CaCO_3$ occurs as an ocean floor sediment:

$$CaCO_3 + H^+ \rightarrow Ca^{2+} + HCO_3^-$$

The addition of hydroxide ions (OH^-) is countered by

$$OH^- + HCO_3^- \rightleftharpoons CO_3^{2-} + H_2O$$

and

$$OH^- + H_2CO_3 \rightleftharpoons HCO_3^- + H_2O$$

(from Gymer 1973, p. 362). The carbonate in the form of calcium carbonate may be precipitated out if the sea water is near saturation.

The ultimate control on this buffering system is the input of dissolved carbon dioxide and the presence of solid calcium carbonate in sediments. An equilibrium situation will be achieved between the amount of CO_2 in the atmosphere and in the ocean. The tendency is for the amount of CO_2 in the atmosphere to increase because of the ever accelerating usage of fossil fuels. But the increase in atmospheric CO_2 has been much reduced because the oceans have acted as a huge reservoir for the extra CO_2—in effect buffering processes have operated to counteract the increase of H^+ ions. For optimum buffering, there needs to be oceanic mixing so that the surface layer of water, high in dissolved CO_2, is circulated to lower depths where the carbonate sediments are present. Thus efficient mixing of the upper and lower layers of the ocean is necessary in order that buffering processes are optimized; however, such mixing does not occur very quickly so that there is a considerable time lag between the absorption of increased quantities of atmospheric CO_2 and ultimate attainment of equilibrium. This example demonstrates the nature of buffering processes; another well quoted example is human blood which is maintained very near to 7·4 pH.

5.3 Solution

A solution is a homogeneous material consisting of two or more pure substances. In a strict sense there can be solid, liquid or gaseous solutions. For example a solution can be obtained by dissolving one gas in another or by dissolving any solid, liquid or gas in a liquid. It is also possible to have solid solutions such as alloys whereby atoms of one metal are dispersed through those of another. In a strict sense water itself is a solution since it is a mixture of H^+ and OH^- ions and H_2O molecules. The term soil solution is used to stress that water in soil contains a large variety of ions, many of which are required as nutrients by plants. This solution eventually reaches streams so that the water in rivers is a complex solution.

Some measures of solution concentration were introduced in chapter 2; a mole is the fixed number of molecules ($6·022 \times 10^{23}$) in the gram molecular weight of a compound and the molarity (M) of a solution is the number of moles of a solute in one litre of solution. Suppose a $1·0M$ solution of copper sulphate ($CuSO_4$) was required, then ($63·5 + 32·1 + 16·0 \times 4$) g of $CuSO_4$ would be dissolved in water and made up to one litre. A similar measure is molality (m) which is the number of moles of a solute in 100 g of a solvent. So to make a $1·0m$ solution of $CuSO_4$, 159·6 g are dissolved in 1000 g of water. In a strict sense, the concentration should be expressed in terms of $CuSO_4$ molecules, and Cu^{2+} and SO_4^{2-} ions because of dissociation. Another measure of solution concentration is the percentage of a solute. For example, a 10% solution of hydrogen peroxide (H_2O_2) by mass means that 10 g of H_2O_2 is dissolved in every 100 g of solution. An older measure of concentration is *normality* though its use is decreasing. Before this measure can be defined, the nature of an *equivalent* needs to be introduced. The equivalent of an acid is the number of grams of the acid which produces one gram of hydrogen ions. Consider the dissociation of hydrochloric acid:

$$HCl \rightleftharpoons H^+ + Cl^-$$

The gram molecular weight of HCl is 36·5 g and this yields 1 g of H^+ ions according to this equation. The equivalent of HCl is thus 36·5 g and when this is made up to a solution of 1 litre, the solution is said to be normal (1N). As the dissociation of carbonic acid is

$$H_2CO_3 \rightleftharpoons 2H^+ + CO_3^{2-}$$

then the gram molecular weight of H_2CO_3 (62 g) needs to be divided by two in order to obtain the equivalent. The equivalent of a base is the number of grams of it which react with one gram of hydrogen ions. Calculations can be carried out in a similar way to the acid examples to obtain equivalents for particular bases and thus solutions of these bases can be made up to particular normalities assuming the bases are sufficiently soluble.

The actual process of solution requires some consideration since this helps to explain the factors influencing solubility. In chapter 4 the nature of entropy was examined and it was shown that the entropy of a system can be thought of as the degree of disorder. Imagine two beakers of equal volume, one containing water and the other vinegar (acetic acid); the sum of the entropies of these two beakers is less than the resultant entropy when the two volumes are mixed in one beaker. This can be understood by noting that there are more possible molecular configurations in the mixture than in the two separate beakers, thus entropy has increased. The usual consequence of solution is the migration of the system to greater entropy and lower energy though it is possible for solution to lead to states of high energy. To be more specific, for solution to occur, there must be a decrease in free energy which is obtained from the equation

$$\Delta G = \Delta H - T\Delta S \quad \text{(equation (4.14) of chapter 4)}$$

where ΔG is the change in free energy, ΔH the change in enthalpy, T absolute temperature and ΔS, the change in entropy. As described in the previous chapter, a process occurs spontaneously if there is a decrease in G, i.e. ΔG has a negative value. The water and vinegar example demonstrated that ΔS for solution is always positive and thus a negative ΔG is obtained by a negative ΔH and a positive ΔS. Solution cannot occur if ΔH is positive and cannot be offset by ΔS unless very high temperatures exist. The practical expression of energy changes associated with the process of solution is often the change in temperature. As previously explained, if ΔH is negative, the process is exothermic. For example when concentrated sulphuric acid is mixed with water, much heat is liberated and this is the reason why acid should be diluted by pouring it into water rather than vice versa. The other possibility, if $\Delta H \neq O$, is for ΔH to be positive, then an endothermic process occurs such as results when potassium chloride is dissolved in water—the temperature drops.

It has been implicit in the preceding discussion that there is an upper limit to the amount of solute which can be dissolved in a quantity of solvent—this limit is the saturated solution. As a simple example, more and more common salt can be added to a beaker of water until no more can be dissolved; when this saturated situation is achieved, the rate of solution of salt grains will be perfectly matched by the rate of precipitation out of solution. The concentration of a saturated solution, a consequence of the solubility, is dependent upon the solvent, solute and temperature. It seems to be the case that if a solute and solvent are composed of similar molecules in terms of structure and electrical properties, then high solubility is favoured. For example, water molecules display polarity (different electrical charges at opposite ends) and thus water is an excellent solvent for other polar solutes—ionic solids. In contrast a non-polar solvent such as carbon tetrachloride (CCl_4) may be used for removing stains of grease or fat. The effect of temperature on solubility depends on the particular solute and solvent, but if the process is endothermic, then solubility will increase with temperature; in contrast if the process is exothermic, then solubility will decrease as temperature increases. When gases are dissolved in water, it is common for their solubility to decrease as temperature increases though this does not apply when solvents other than water are used. Such apparent variability in behaviour can only be understood by reference to the energetics of the solution—in particular to equation (4.14).

The qualitative nature of solution has been described, but it is now necessary to express quantitatively the solubility of different substances. Consider the

Table 5.2 Solubility products for selected compounds (from Gymer 1973, p. 278).

Compound	Formula	Solubility product (K_{sp})
Calcium carbonate	$CaCO_3$	$4{\cdot}7 \times 10^{-9}$
Calcium hydroxide	$Ca(OH)_2$	$5{\cdot}5 \times 10^{-6}$
Calcium phosphate	$Ca_3(PO_4)_2$	$2{\cdot}0 \times 10^{-29}$
Calcium sulphate	$CaSO_4$	$2{\cdot}4 \times 10^{-5}$
Iron (II) carbonate	$FeCO_3$	$3{\cdot}5 \times 10^{-11}$
Iron (III) phosphate	$FePO_4$	$1{\cdot}3 \times 10^{-22}$
Lead carbonate	$PbCO_3$	$1{\cdot}5 \times 10^{-13}$
Lead sulphate	$PbSO_4$	$1{\cdot}7 \times 10^{-8}$
Magnesium carbonate	$MgCO_3$	$1{\cdot}0 \times 10^{-5}$
Magnesium phosphate	$Mg_3(PO_4)_2$	$\sim 1 \times 10^{-27}$
Manganese (II) carbonate	$MnCO_3$	$8{\cdot}8 \times 10^{-11}$
Silver chloride	$AgCl$	$1{\cdot}6 \times 10^{-10}$

ionic solid, potassium chloride, KCl. Saturation occurs when there is still solid KCl present in the solution. At equilibrium

$$KCl(s) \rightleftharpoons K^+ + Cl^-$$

where s indicates the solid state. The *solubility product* (K_{sp}) is obtained by multiplying the molar concentrations of K^+ and Cl^-, i.e.

$$K_{sp} = [K^+][Cl^-]$$

Suppose the solution of magnesium phosphate $Mg_3 (PO_4)_2$ is considered:

$$Mg_3 (PO_4)_2 (s) \rightleftharpoons 3Mg^{2-} + 2PO_4^{3-}$$

In this case

$$K_{sp} = [Mg^{2-}]^3 [PO_4^{3-}]^2$$

The solubility products for selected compounds are presented in table 5.2. The values of the solubility products assume that the individual substances are dissolved in pure water. If on the other hand a slightly soluble substance is dissolved in a solution, including a type of ion the same as in the substance, then a smaller quantity of the substance is required to reach saturated conditions. Such a situation is very applicable to the physical environment since substances are never dissolved in pure water. It is also relevant to mention possible consequences when two solutions are mixed. For example, precipitation could occur should the resultant solution prove to be supersaturated. Gymer (1973) considers a mixture of solutions of silver nitrate ($AgNO_3$) and sodium chloride (NaCl). It is possible for the Ag^+ and Cl^- ions to precipitate out of solution as AgCl. This will occur if the ion product obtained by calculating $[Ag^+][Cl^-]$ for the actual solution is greater than the saturation product (K_{sp}) for AgCl. If the ion product is less than K_{sp}, then the solution is unsaturated with respect to the Ag^+ and Cl^- ions.

5.4 Oxidation–reduction

Solution involves no transfer of electrons; in contrast, reactions in which electrons are transferred between atoms are known as oxidation–reduction or redox reduction. The substance which loses electrons is oxidized whilst the substance with the corresponding gain is reduced. Redox processes are widespread in the natural environment; for example the characteristic brown colour of the B horizon of well drained temperate soils results from the oxidation of iron.

The simplest approach to oxidation is to view it as a process whereby oxygen is added to a substance; similarly reduction is the loss of oxygen. The oxidation of iron results in either iron (II) oxide or iron (III) oxide—traditionally called ferrous oxide (FeO) and ferric oxide (Fe_2O_3). In these oxides, iron displays different valencies—two and three respectively. The difference between these two oxides will be discussed later in this section; suffice it to state for the moment that Fe can be oxidized first to FeO and then further to Fe_2O_3. If either of these oxides is added to hydrochloric acid, the corresponding chlorides are formed. These changes can be expressed as follows:

$$Fe_2O_3 \xrightarrow{HCl} FeCl_3$$

$$(1) \uparrow O_2 \qquad Cl_2 \uparrow \ (2)$$

$$FeO \xrightarrow[HCl]{} FeCl_2$$

(from Stamper and Stamper 1971). Oxidation is clearly taking place in reaction 1. Reaction 2 is obtained by the effect of the gas chlorine on iron (II) chloride and since it parallels reaction 1 is also an oxidation process. This implies that oxidation can mean the addition of any non-metallic element (not hydrogen) to a compound.

Common oxidizing agents are oxygen gas, chlorine gas, potassium permanganate ($KMnO_4$) and potassium dichromate ($K_2Cr_2O_7$). For example the latter substance is used in analysis to determine the amount of organic matter in soils; the amount of potassium dichromate necessary to oxidize the organic matter is found by quantitative analysis and thus the organic content can be predicted. An oxidizing agent of particular importance in the atmosphere is the gas ozone (O_3). This triatomic oxygen is produced in the stratosphere by the action of ultraviolet radiation on ordinary oxygen: two oxygen atoms are produced from each molecule and then one oxygen atom combines with an oxygen molecule. The reaction is limited in occurrence and is most marked at a height of 27 to 30 km near the equator. The ozone is transported to high latitudes as well as to lower altitudes. Besides its ability to absorb radiation, ozone despite its occurrence in very small quantities plays a natural cleansing action in the atmosphere because of its oxidizing ability.

Oxidation as the process of electron transfer can be exemplified by considering the fusion of one atom of sodium with one of chlorine to form a molecule of sodium chloride. Sodium is $^{23}_{11}Na$ whilst chlorine is $^{35}_{17}Cl$. The Na atom thus has 11 electrons, 2 in the first shell, 8 in the second and 1 in the third and in a reaction whereby Na changes to a more stable state, this outer single electron will be removed. Similarly the chlorine atom has shells of 2, 8 and 7 electrons, one short in the outermost shell before it is complete. The reaction means that the excess electron from the Na atom is removed (oxidized) whilst the Cl atom gains one electron (reduced). The convention is to say that sodium has an oxidation state (or number) of $+1$ whilst chlorine has a corresponding value of -1 in the compound NaCl. These values describe the charge which an atom seems to have in a compound and can be determined by consideration of electronic configuration. Rules have been devised to permit the quick determination of oxidation states since their use helps in an understanding of the process as well as allowing the balancing of redox equations.

Some of the rules are as follows:

(1) The oxidation states of elements which occur free in nature are 0. For example atoms of oxygen occurring as the gas O_2 have an oxidation number of 0.
(2) The oxidation states of alkali metals (group I of the periodic table, viz. lithium, sodium, potassium, rubidium, cesium and francium) in their compounds are $+1$; the corresponding value for alkaline earths (group II of the periodic table, viz. beryllium, magnesium, calcium, strontium, barium and radium) is -1. Thus the value of $+1$ for Na in NaCl accords with this rule.
(3) The oxidation state for oxygen in compounds is usually -2 and similarly for hydrogen in compounds the value is $+1$. Exceptions occur respectively with peroxides and hydrides.
(4) In a molecule the sum of the oxidation states must equal zero, whilst with an ion, the sum must equal the net charge.

As an example take the oxidation state of sulphur in sulphuric acid (H_2SO_4). Hydrogen has the oxidation state $+1$ whilst oxygen has the state -2.

$$(H^{+1})_2 S^? (O^{-2})_4$$

In order for there to be a balance, S in H_2SO_4 must have an oxidation state of $+6$. (Rules modified from Anderson, Ford and Kennedy (1973), Sienko and Plane (1976).)

Oxidation occurs when there is an increase in oxidation state and similarly reduction is reflected in a decrease of state. The oxidizing agent is the substance which causes the oxidation and at the same time is itself reduced and vice versa for a reducing agent. In the example of NaCl, before fusion both elements have states of 0 (rule 1), but after, the Na is oxidized whilst Cl is reduced. It must not be assumed that oxidation states for particular elements are constant. For example iron has the states $+2$ and $+3$, copper $+1$ and $+2$, and tin $+2$ and $+4$.

As already hinted the oxidation of Fe (II) to Fe (III) is particularly evident in many soils. Iron exists in many rock minerals in the lower oxidation state; it can be oxidized to the higher state while the iron is still integral to the crystal structure thus inducing crystal disintegration; alternatively the iron, on release from the crystal, may be in the iron (II) state, but is immediately oxidized to the iron (III) state. For example, the mineral olivine on reaction with water yields another mineral called serpentine as well as silica and ferrous oxide which is immediately oxidized to iron (III) oxide (haematite):

$$3MgFeSiO_4 + 2H_2O \rightarrow H_4Mg_3Si_2O_9 + SiO_2 + 3FeO \qquad (5.20)$$

Olivine   Serpentine   Silica   Iron (II) Oxide

$$4FeO + O_2 \rightarrow 2Fe_2O_3 \qquad (5.21)$$

Iron (III) Oxide

(after Brady 1974).

Compounds which absorb water are usually formed rather than iron (III) oxide, for example goethite ($Fe_2O_3 . H_2O$) which is reddish-brown and is one of the main colouring constituents of soils. The attachment of water molecules in this way is called hydration. The oxidation of iron (II) oxide to iron (III) oxide is only possible if the soil is well aerated; with poor drainage conditions the iron remains in the reduced state reflected in the greyish-blue colour of waterlogged soils.

An association between ionization and oxidation should be apparent— both are connected with electron change. For a gas the ionization energy is the energy required to remove an outermost electron; this depends upon the distance of the electron from the nucleus as well as the charge of the nucleus. In a solution ions are attracted to water molecules because of their polarity; again the attraction depends upon the size of the ion and its charge. The attraction in this situation is called the ionic potential, the weathering significance of which is outlined by Ollier (1969). A related concept is redox potential which expresses the relative stability of particular oxidation states. Such stability is dependent upon the energy required to add or remove electrons; again Ollier (1969) illustrates the significance of this potential.

5.5 Hydration and hydrolysis

Given the importance of hydration and hydrolysis in weathering, it would seem appropriate to illustrate the more particular nature of these processes. Hydration occurs when water is absorbed to form a new compound which is little different from the original one; an example is the hydration of haematite to limonite, viz.

$$2Fe_2O_3 + 3H_2O \rightarrow 2Fe_2O_3 . 3H_2O \qquad (5.22)$$

Haematite   Limonite

Haematite is bright red and causes the distinctive colour of soils in tropical and subtropical areas. When hydration occurs, the colour changes to yellow (limonite), but this can easily be reversed if dehydration occurs. Few minerals undergo direct hydration, but mica is a noted exception since some H^+ and OH^- ions are able to penetrate in between the plate-like layers resulting in expansion and physical degeneration. For weathering, hydration tends to be considered a secondary process since it often follows hydrolysis which is probably the most important chemical weathering process (Fitzpatrick 1971, Brady 1974). In contrast to hydration, hydrolysis involves chemical change whereby cations are replaced by hydrogen ions. The exact nature of hydrolysis varies from mineral to mineral, but it seems that basic cations (e.g. Na, K, Ca and Mg) are always exchanged for hydrogen ions. This can be demonstrated by the hydrolysis of the mineral microcline:

$$KAlSi_3O_8 + HOH \rightarrow HAlSi_3O_8 + KOH \qquad (5.23)$$

Microcline   Aluminosilicic
acid

The potassium hydroxide then reacts with carbonic acid to yield potassium carbonate which is soluble in water. The aluminosilicic acid also undergoes chemical change producing aluminium and silicon compounds:

$$2HAlSi_3O_8 + 8HOH \rightarrow Al_2O_3 . 3H_2O + 6H_2SiO_3 \qquad (5.24)$$

The initial exchange of potassium ions for hydrogen ones (equation (5.23)) has little effect on the structure of the microcline, but the subsequent removal of aluminium (equation (5.24)) causes the disintegration of the original mineral. The rate of hydrolysis is influenced by the concentration of hydrogen ions— the lower the pH, the faster the rate. Hydrolysis necessarily means the consumption of hydrogen ions leading to the formation of a basic solution. Hydrolysis in soils is also aided by the presence of chelating agents which are formed by biological processes: these agents, which are excreted by lichens, have the ability to extract ions from otherwise insoluble solids and thus enable the exchange of ions (Ollier 1969). The presence of chelating agents assists the process of hydrolysis, but chelation itself, which describes the incorporation of an ion into an organic ring molecular structure is highly significant and Birkeland (1974) has suggested that in certain situations may be more important than hydrolysis. In concluding this section it is worth stressing the interdependence of reactions. The weathering of any one mineral, for example, may well involve hydration, solution and hydrolysis.

5.6 Weathering and thermochemistry

To conclude this chapter it is appropriate that consideration is given to the energetics of chemical processes since this helps to explain why minerals weather at various rates. It has for long been recognized that the resistance of a mineral to weathering depends upon such characteristics as size of grain, surface area, hardness, cleavage, solubility and the nature of its atomic structure. Goldich (1938) put forward a ranking of minerals with the least stable at the top and most stable at the bottom:

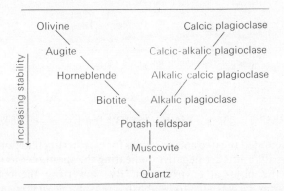

This implies that in a given locality olivine will weather more rapidly than augite, etc. Such an empirical approach accords with the frequent absolute dominance of quartz in soils. A similar sequence of minerals to the above was proposed by Bowen (1922) in a rather different argument; he was concerned with the evolution of igneous rocks and worked out the order of crystallization of many common minerals. The resultant Bowen reaction series thus expresses the order in which the different minerals begin to crystallize during the solidification of basaltic magna. For example, olivine and calcium rich plagioclase are the first minerals to crystallize in abundance from molten basalt. An explanation for the parallel between the Bowen reaction series and

Goldich's stability series has been the focus for much research, but the trend in recent years has been towards a thermochemical approach. Curtis (1976a, b) considers the energy changes which occur in weathering reactions. In chapter 4 the following equation was introduced and exemplified:

$$\Delta G^\circ = \sum \Delta G^\circ_{f\,(products)} - \sum \Delta G^\circ_{f\,(reactants)}$$

where ΔG° is the free energy of the reaction, and ΔG°_f the free energy of formation of either the products or reactants. If ΔG° has a value greater than 0, the reaction does not occur at the standard temperature; if the value is negative, then the reaction is spontaneous. Thus the outcome for mixing minerals with other potential reactants can be predicted if all the free energies of formation are known. Curtis has drawn together many equations of chemical weathering with estimates of free energies of formation for many minerals.

As an example, consider the weathering of the mineral wollastonite, which is a member of the zeolite group:

$$CaSiO_3 + 2H^+ \rightarrow Ca^{++} + H_2O + SiO_2$$

Wollastonite   Silica

$$\Delta G^\circ = \Delta G^\circ_f(Ca^{++}) + \Delta G^\circ_f(H_2O) + \Delta G^\circ_f(SiO_2) - \Delta G^\circ_f(CaSiO_3) - \Delta G^\circ_f(2H^+)$$

The relevant free energies of formation are

		$\Delta G^\circ_f (kJ\ mol^{-1})$
Ca^{2+}	(aq)	$-553 \cdot 1$
H_2O	(l)	$-237 \cdot 2$
SiO_2	(s)	$-856 \cdot 5$
$CaSiO_3$	(s)	$-1549 \cdot 3$
H^+	(aq)	0

(adapted values from Curtis 1976).

Thus

$$\Delta G^\circ_f = (-553 \cdot 1 - 237 \cdot 2 - 856 \cdot 5 + 1549 \cdot 3)kJ mol^{-1} = -97 \cdot 5\ kJ\ mol^{-1}$$

Curtis has calculated the free energy values for weathering reactions of twenty-four minerals; all results are negative, indicating the spontaneous nature of the reactions in the physical environment. By converting the results into a comparable form, he was able to establish a reasonable degree of correlation between energy and persistence in weathering.

In no way has this chapter attempted to present a comprehensive description of specific types of chemical processes. Instead the aim has been to introduce the principles behind these processes with the hope that the reader gains an outline theoretical framework onto which he can attach any chemical process in the physical environment. The examples have been selected primarily from weathering processes, but atmospheric or biological reactions could have been used. The last part of the chapter illustrates once again that an energy approach deepens an understanding of processes in the physical environment.

6
Heat and phase change

In the last part of the previous chapter the transfer of energy through chemical reactions was illustrated with reference to weathering. The ultimate source of energy for all processes in the physical environment is the sun; energy is transmitted to the earth's surface in the form of radiation which warms the surface. A temperature gradient is thus established down the first few metres of the earth's crust leading to the downward transfer of energy in the form of heat, a process which is arrested as soon as uniform temperatures are encountered. Temperature differences are also established over the surface of the earth causing heat transfer by atmospheric circulation. Similar redistribution of energy occurs in the oceans. Inputs of heat also change the states of substances—in the physical environment the most important substance is water and it occurs widely in its three states or phases, solid, liquid and gas. In this chapter attention is focused on the nature and consequences of energy transfer in the form of heat.

6.1 Equilibrium, expansion and heat capacity

To take the temperature of water in a beaker a thermometer is inserted; heat is transferred through the glass bulb causing expansion or contraction of the mercury. A constant reading implies that the thermometer is in thermal equilibrium with its surroundings. This introduces the fundamental nature of temperature since the temperature for any system will establish whether or not the system is in thermal equilibrium with any other. Obviously when two adjacent systems are in thermal equilibrium (no net heat transfer), they will have the same temperature. The nature of the Celsius and Kelvin temperature scales has already been outlined (chapter 1).

Nearly all solids, liquids and gases, on the application of heat, expand. A noted exception is water which decreases in volume between 0 °C and 4 °C. The general increase in volume is a reflection of the increased agitation of individual atoms or molecules meaning that they move slightly further apart. For the moment, consideration is limited to solids and liquids since pressure–temperature–volume relationships of gases are discussed in chapter 7. The expansion of solids can be expressed in a linear, areal or volumetric manner though the expansion of liquids can only be stated in volumetric terms. A metal rod of length L, for example, when subjected to a temperature change ΔT by being heated will suffer a change in length ΔL. Thus the coefficient of linear expansion (α) can be defined by

$$\alpha = \frac{\Delta L}{L \Delta T} \qquad (6.1)$$

the change in length per unit length and per unit temperature change. The values of α also depend on the particular temperature range since the rate of increase of length tends to be larger at high-temperature ranges. Metals clearly expand and contract at different rates; if two contrasting metals are fixed together to form a bimetallic strip, an increase in temperature will cause one side to expand more than the other so that the strip will bend. This phenomenon is utilized in thermostats. The coefficient of areal expansion for solids is defined as the increase in area per unit increase of temperature; this quantity is not widely used. Of much greater significance is the coefficient of volume expansion (β), applicable to solids and liquids. This can be defined as follows:

$$\beta = \frac{\Delta V}{V \Delta T} \qquad (6.2)$$

where V is the original volume, ΔV the increase in volume and ΔT the temperature change. Again this coefficient varies a little according to the temperature range. For solids, the coefficient of volume expansion is three times that of the linear value. Volume changes resulting from temperature variations have consequences with regard to physical weathering. Rocks tend to be poor conductors of heat so that a temperature difference can exist between the surface of a rock and a short distance inside. The surface skin thus expands more than the interior of the rock creating various stresses within the rock so that fracture is possible. As rocks are made up of various minerals which may warm up at different rates and may also have different coefficients of volumetric expansion, then the consequence could be physical weathering. The efficiency of these processes in terms of weathering has been questioned, but Ollier (1969), in reviewing the topic, describes examples from Australia where thermal changes have caused weathering.

Since volume changes with temperature, densities of substances must vary similarly. Values for water are presented in table 6.1; note the maximum density at 4 °C.

Table 6.1 Density of water at various temperatures (from Weast 1974, F-11).

Temperature (°C)	Density (g cm^{-3})
0	0·9999
3·98	1·0000
10	0·9997
20	0·9982
50	0·9881
100	0·9584

When two bodies at different temperatures are brought together, the ultimate thermal equilibrium state will be characterized by a temperature dependent not only upon the original temperatures of the two bodies, but also upon their thermal natures. So there is another thermal property additional to temperature and this is called *thermal capacity*. For example, consider several substances of equal volume, all at the same temperature; these substances will require varying quantities of heat to elevate them to another temperature because of their different thermal capacities. In fact the heat capacity of any one of the substances can be determined by dividing the amount of heat supplied (ΔQ) by the temperature change (ΔT). For comparative purposes it is more convenient if the heat capacity is expressed per unit mass of the substance and thus *specific heat capacity* (or just *specific heat*) is defined as follows:

$$C = \frac{\Delta Q}{m\Delta T} \qquad (6.3)$$

where C is the specific heat capacity, ΔQ is supplied heat, ΔT is the temperature change and m is the mass. A high specific heat capacity indicates that a substance, when heated, will not change in temperature as markedly as one with a low specific heat capacity. This can be demonstrated by calculating the temperature change of $1 \cdot 0$ kg of the substances in table 6.2 given their specific heat capacities.

Table 6.2 Specific heat capacities of selected substances and associated temperature changes on absorption of 4·2 kJ (adapted from Townsend 1973, p. 132).

Substance	Specific heat capacity (kJ kg^{-1} K^{-1})	Increase in temperature of 1 kg on absorption of 4·2 kJ (K)
Water	4·2	1·0
Quartz sand	0·8	5·3
Kaolinite clay	1·0	4·2
Humus	1·9	2·2

These changes in temperature are obtained by appropriate substitution of values determined by experiment into equation (6.3). It should be stressed that the values given in table 6.2 are only averages; for example variations in C for quartz sand are to be expected due to differences in grain size and packing. Sometimes the mass term in this equation is replaced by the number of moles, obtained by dividing the mass by the molecular weight of the substance. The resultant thermal property is known as the *molar heat capacity*. Specific heat capacity has units such as kilojoules per kilogram per degree Kelvin (table 6.2) whilst molar heat capacity has units of joules per mole per degree Kelvin. For example, for water,

$$\text{molar heat capacity} = 75 \cdot 4 \text{ J mol}^{-1} \text{ K}^{-1}$$

Equation (6.3) states that to determine specific heat capacities, it is necessary to measure the effect on temperature of the input of a certain heat quantity per unit mass. One laboratory procedure is to supply the heat electrically and to monitor the input and output of electrical energy. Temperature changes can be recorded by using thermistors which are small electrical resistors which vary in resistance according to temperature. Over common temperatures specific heat capacities can be considered to be constant for most practical purposes; however, these capacities decrease as temperature drops. The form of specific heat capacity/temperature curves is the subject of much modern research in solid-state physics since theory linking atomic structure, the nature of energy within atoms and within lattice structure, and heat capacity can be postulated.

Brief consideration of soil thermal regimes provides a practical example of heat capacity. The input of heat to soil is from direct and diffuse solar radiation

D*

which of course varies according to time and place. The response of the soil at the surface to this input depends upon the specific heat capacity of the soil and this varies according to the nature of the soil. Light sandy soils warm up more rapidly than say heavier loams or clays and this can be important for market gardening. For soil thermal studies it has been found useful to determine heat capacity on a volumetric rather than mass basis, so the volumetric heat capacity (C_v) is the amount of heat needed to change the temperature of a unit volume by one degree. Assuming unfrozen soil, this volumetric heat capacity is dependent upon the dry bulk density of the soil, its moisture content and the specific heat capacity of the constituent mineral matter (Jumikis 1966). Jumikis specifies the relationship as

$$C_v = \gamma_d(C_s + C_w W/100) \tag{6.4}$$

where, after modification to conform to the SI system, C_v is the volumetric heat capacity (J cm^{-3} K^{-1}), γ_d is the dry bulk density (g cm^{-3}), C_s is the specific heat capacity of dry mineral matter (J g^{-1} K^{-1}), C_w is the specific heat capacity of water (J g^{-1} K^{-1}) and W is the moisture content on a percentage mass basis. Equation (6.4) states that the volumetric heat capacity of soils increases in response to increases in bulk density and moisture content. Joynt (1973) has computed values of C_v for normal ranges of moisture content, bulk density and specific heat capacity of dry mineral matter. Later in this chapter the use of volumetric heat capacities is illustrated (section 6.4), but demonstration is only possible after some consideration is given to the transfer of heat.

6.2 Transfer of heat

So far consideration has been limited to the static situation whereby the earth's surface warms up in response to the receipt of radiation. Of course the warming of the surface creates a temperature gradient in the soil and heat will be transferred downwards. The form of the temperature/depth curve depends upon the rate of warming of the surface, variations in soil volumetric heat capacity and the ability of the soil to transmit heat. The transfer of energy through the soil is achieved primarily by conduction. Other major heat transfer processes are convection and radiation.

(1) Conduction

Any metal rod, if heated at one end, will conduct heat along to the other end. Conduction is the transference of heat from particle to particle within a substance. Suppose that the rod is perfectly insulated from the surrounding environment and that one end is maintained at temperature T_1; heat is transferred along to the other end and after some time the temperature at the cooler end will be constant—say T_2. The temperature gradient over the length of the rod (L) is thus $(T_1 - T_2)/L$. Intuitive reasoning suggests that the rate of flow of heat (H) along the rod will increase as the temperature difference increases, will increase as the cross-sectional area (A) of the rod increases and will be inversely proportional to the rod length (L). Thus

$$H \propto \frac{A(T_1 - T_2)}{L}$$

The heat conducting ability of the rod must also be taken into account and thus the thermal conductivity (k) of the rod is defined as the constant of proportionality. Thus

$$H = \frac{kA(T_1 - T_2)}{L} \tag{6.5}$$

This can be re-arranged as

$$k = HL/[A(T_1 - T_2)] \tag{6.6}$$

so values of k can be obtained by experimentation. Thermal conductivity is expressed in watts per metre per degree Kelvin, $W\,(mK)^{-1}$. As a reminder, one watt is a measure of power—the utilization of one joule per second.

In terms of heat transfer in rock or soils, the assumption tends to be made that such transfer takes place only by conduction. For material which is not frozen, thermal conductivity depends on the orientation and composition of the aggregate and on the size, orientation and moisture content of the pores (Judge 1973). A variety of methods for determining thermal conductivity are in use (Taylor and Jackson 1965). For example, Joynt (1973) measured thermal conductivity by monitoring temperature gradient away from a line source of heat embedded in the samples. Some thermal conductivities of common rock types are presented in table 6.3.

Table 6.3 Thermal conductivities of selected rock types (from Judge 1973, p. 100).

Rock type	Thermal conductivity $(W\,m^{-1}\,K^{-1})$
Shale	1·5
Limestone	2·9
Sandstone	4·2
Gneiss	2·5
Quartzite	5·9
Gabbro	2·5
Granite	2·9

These values indicate the variability in the thermal conductivity of rocks; quartzite can transmit heat almost four times as fast as shale. The thermal conductivity of ice is about four times that of water, so that the effect of freezing of rock or soil is to increase its conductivity.

Thermal conductivity (k) and volumetric heat capacity (C_v) have been defined and when the former is divided by the latter, the *diffusivity* (D) is obtained. Thus

$$D = k/C_v$$

The units of diffusivity are worked out as follows:

$$D = \frac{k}{C_v} = \frac{W\,m^{-1}\,K^{-1}}{J\,m^{-3}\,K^{-1}} = m^2\,s^{-1}$$

The nature of diffusion can be demonstrated by visualizing the diffusion of heat through a soil. Suppose that there is a temperature gradient in the soil so that at an upper level the temperature is higher than at a lower level. Let the thickness between these two levels be h and the area under consideration A.

Also suppose that the heat at the upper level can be expressed in terms of energy per unit mass (Q_1) whilst Q_2 represents the corresponding value for the lower level. Then the quantity of heat (Q) which diffuses through A in time t is obtained from

$$Q = DA\frac{(Q_1 - Q_2)t}{h} \tag{6.7}$$

where D is the coefficient of diffusion.

The same approach can be applied to the diffusion of solutions of different concentrations. In thermal problems, measures of diffusivity are often used (for example, see Sellers 1965).

(2) Convection

Heat transfer by convection is achieved by the actual motion of the material. In meteorology convection is limited to describing vertical movement whilst horizontal motion is called advection. The reason for convection is variation in density of the fluid resulting from temperature differences. Hot air rises and cool air descends because of density differences. Consider what happens in a pond at say 10 °C as it cools. Cooling proceeds fastest at the surface leading to water of a slightly higher density than below; this surface water then sinks to be replaced from below by slightly warmer water. A convection cell is thus established and continues until all the water is at 4 °C. Water near the surface then cools to below 4 °C but remains there because the density of water is at a maximum at 4 °C. Thus no convection is possible below 4 °C; the surface then loses heat only by conduction. Water has a very low thermal conductivity and so cooling below 4 °C proceeds slowly.

The quantification of conduction is comparatively easy, but the same is not the case with convection. Complications arise because the amount of heat transferred by convection is conditioned by a large number of factors. Various empirical procedures have been devised using a convection coefficient to tackle such problems as heat loss from buildings, and the interested reader can follow this up in a physics text, such as Sears, Zemansky and Young (1974, pp. 271-2). The various factors influencing the rate of convection in the atmosphere are the basis to an understanding of adiabatic lapse rates (see section 7.2).

(3) Radiation

Radiation is a very different process to conduction and convection since it can operate through empty space as no molecular contact or mass movement is necessary. The process of radiation permits the transfer of energy and thus radiation is expressed in energy units per unit area per unit of time: $J\,m^{-2}\,s^{-1}$ or $W\,m^{-2}$ in the SI system. All bodies, irrespective of their temperatures, emit radiant energy which is in the form of electromagnetic waves, the nature of which was described in chapter 2. If one body is at a higher temperature than an adjacent one, then the latter will absorb radiant energy from the hotter body until both are at the same temperature. Thus it is clear that the rate of emission of radiant energy and the associated wavelengths depend on the temperature of the body—in fact the rate of emission is proportional to the fourth power of the absolute temperature. However, the rate of radiation from a body is also influenced by the nature of the substance and its surface. As a general rule bodies with matt black surfaces are much better radiators than ones with glossy light surfaces. The same also applies with absorption of radiant energy in that matt dark surfaces permit better absorption than glossy light ones.

A body at low temperature will emit little radiation and this will be of a long-wave nature. As the temperature rises, the amount of radiation emitted increases markedly and is in the form of shorter wavelengths. This can be demonstrated by heating a metal rod in a flame; at first no change in colour of the rod is apparent but as it becomes hot it begins to glow because the wavelength of emission has come into the visible part of the spectrum. The variation in wavelength and rate of emission with temperature is illustrated in figure 6.1. The area between the curve and the horizontal axis indicates the total rate of radiation at specific temperatures. The set of curves demonstrates that the rate rapidly increases with temperature and that emission at higher temperatures is in the form of short electromagnetic waves. Figure 6.2 shows the spectral distribution of radiation from the sun and from the earth. The set of curves in figure 6.1 assumes that the radiation is being emitted from a *blackbody* which means that it is a perfect radiator. Similarly if a body absorbs all radiation incident upon its surface, then it is also a blackbody. In practice it is impossible to make perfect blackbodies, though this ideal can be approached by making a small opening into a cavity in a wall; this cavity has a rough internal surface (figure 6.3). Radiation which enters the cavity is partly reflected from point to point within the cavity and partly absorbed by the walls so that there is very little chance of it escaping from the cavity again.

As well as conceiving of a body which absorbs all radiation as in figure 6.3, it is also possible to visualize a body which displays perfect emission of radiation. Of course no such body is known to exist, but such a notion permits

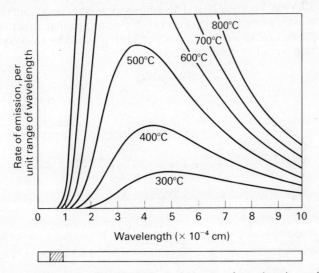

Fig. 6.1 Rate of emission of radiant energy per unit range of wavelength as a function of wavelength. The shaded area indicates the visible part of the spectrum. (from Sears, Zemansky and Young 1974, p. 273)

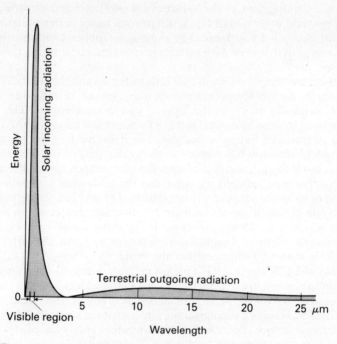

Fig. 6.2 The spectral distribution of solar and terrestrial radiation. (from G. M. B. Dobson: *Exploring the Atmosphere*, Oxford University Press.

comparison of surfaces in terms of absorption and emission of energy. For example, suppose the radiant emittance (or emissivity) per unit area and per unit time for a blackbody is W_{bb}. This emissivity would result from the perfect absorption of the same quantity of energy. Now imagine that an actual surface is exposed to this radiation, only a fraction of W_{bb} would be absorbed and the same fraction would be emitted. Then

$$W = aW_{bb} \tag{6.8}$$

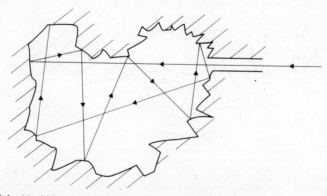

Fig. 6.3 An ideal blackbody formed by a rough cavity in a wall.

where W is the radiative emittance from the surface of any body at a particular temperature, a is the fraction of energy which is absorbed and W_{bb} is the radiant emittance of a blackbody at the same temperature (Sears and Zemansky 1963). A value of $1\cdot0$ for a indicates a perfect absorbing surface whilst at the other extreme, a value of $0\cdot0$ would express perfect reflectivity. The reflectivity or albedo is equal to $(1\cdot0 - a)$. The same argument can be presented for the fraction e, the relative emittance of a surface, so that equation (6.8) can also be written as

$$W = eW_{bb} \qquad\qquad (6.9)$$

The ratio of W/a is the same for *all* surfaces at the same temperatures and is equal to the radiant emittance of a blackbody at that temperature. This statement is known as *Kirchoff's law*.

The rate of energy in watts emitted per square metre per second from a blackbody (W_{bb}) can be determined from *Stefan's law*:

$$W_{bb} = \sigma T^4 \qquad\qquad (6.10)$$

where σ is a constant equal to $5\cdot67 \times 10^{-8}$ W m^{-2} K^{-4} and T is absolute temperature. This is the law expressing the former statement that radiation varies according to the fourth power of the absolute temperature.

In practice no objects or bodies can be considered as perfect blackbodies; bodies only absorb a fraction of the radiation incident upon their surfaces. The absorption factor (a) expresses the fraction of absorbed radiation; so a body with a equal to $0\cdot5$ absorbs only half the radiation incident upon it. Stefan's law can thus be written in the more general form for absorption:

$$W = a\sigma T^4 \qquad\qquad (6.10)$$

The parallel term to express the fraction of radiation which is emitted is the *emissivity* (e). For emission, Stefan's law takes the form

$$W = e\sigma T^4 \qquad\qquad (6.11)$$

A brief illustration of this law can be made with reference to infrared radiation from the earth. Figure 6.2 shows how the earth's surface emits long-wave radiation, much of which is in the infrared part of the electromagnetic spectrum. The term *flux* is used to describe the rate at which energy passes through a unit cross-sectional area in unit time. The terms W in equations (6.9) and (6.10) are examples of fluxes. Suppose it was necessary to estimate the global average flux of emitted infrared radiation, then equation (6.11) could be used as follows:

$$\text{terrestrial flux} = e\sigma T^4$$

where σ is the constant $5\cdot67 \times 10^{-8}$ W m^{-2} K^{-4}, T is absolute temperature (average of about 285 K for earth's surface) and e is an average value for infrared emissivity. Over the earth's surface there are marked variations in the emissivity of infrared radiation, a phenomenon utilized in aerial photography when a film sensitive to infrared radiation is used. Some values of e for various surfaces are presented in table 6.4.

Thus in order to estimate the total average terrestrial flux from emitted infrared radiation, an average value for e would have to be obtained based on

Table 6.4 Selected values of infrared emissivity
(from Sellers 1965, p. 41).

Surface	Infrared emissivity, e
Water	0·92–0·96
Fresh fallen snow	0·82–0·995
Ice	0·96
Dry light sand	0·89–0·90
Wet sand	0·95
Coarse gravel	0·91–0·92
Dry ploughed soil	0·90
Oak woodland	0·90
Pine forest	0·90

variations in surface conditions. Alternatively consideration may be limited to predicting this flux for a much smaller homogeneous area for which better estimates of temperature and emissivity can be made.

6.3 Phase changes

Matter can exist in one of three states—solid, liquid or gas. These states can easily be identified but return has to be made to the kinetic nature of matter in order to understand them. When molecules of a substance are not able to move about within a medium, but can only vibrate about their mean positions, the substance must be a solid. If each molecule or atom has a characteristic location within a regular array, then the solid is crystalline. Where there is no regular structure, the solid is amorphous. In a liquid individual molecules move about freely, but find difficulty in leaving the surface. Thus a greater degree of disorder is present within a liquid than in a solid since there are fewer constraints on individual molecules. Gases have no such constraints and molecules move about randomly. The greatest molecular disorder is thus exhibited by gases. Liquids and gases possess fluidity. Solids and liquids are described as condensed forms of matter indicating that per unit mass they occupy small volumes; in contrast unit masses of gases at standard temperature and pressure, occupy considerably larger volumes. The nature and properties of solids, liquids and gases are considered in detail in the following chapter; for present purposes it is important to note that the three states are characterized by various degrees of disorder or entropy. This implies that the presence of particular phases is conditioned by energy. For example, when heat is applied to ice, it melts and further application of heat causes the water to evaporate to the gaseous state.

Suppose ice at $-25\ °C$ is placed in an insulated container which has an inbuilt coil. The steady application of heat causes the ice to warm up at a constant rate as shown by line AB on figure 6.4. The input of energy means that the atoms in the crystal lattice increase in their individual energies. As soon as the ice reaches 0 °C, melting occurs because the energy of the atoms is sufficiently great to break the bonds of the crystal lattice; even though heat is being applied constantly, there is no further increase in temperature until all the ice has melted (line BC). This phase change thus requires heat and the quantity of heat per unit mass necessary to melt a substance at the same temperature is called the *latent heat of fusion*. This change in enthalpy can be represented by the symbol ΔH_{fus} and the thermochemical change is as follows:

$$H_2O \text{ (s)} \rightarrow H_2O \text{ (l)} \quad \Delta H_{fus} = 333 \cdot 7 \text{ J g}^{-1}$$

This states that the conversion of one gram of ice at 0 °C requires 333·7 J to change it to water at the same temperature. After all the water is in the liquid state, with the continued application of heat, the temperature rises (line CD) resulting in an increase in energy of each molecule. This is because these molecules possess greater and greater kinetic as well as internal energies. The slope of the line CD is less than that of AB because the specific heat of water is greater than that of ice. When 100 °C is achieved, assuming standard pressure conditions, all individual water molecules have sufficient energy to escape from the liquid through vaporization within the liquid though some would have been lost by evaporation at lower temperatures. This phase change again requires heat and this can be represented as follows:

$$H_2O \text{ (l)} \rightarrow H_2O \text{ (g)} \quad \Delta H_{vap} = 2256 \cdot 7 \text{ J g}^{-1}$$

where ΔH_{vap} is called the *latent heat of vaporization*. Line DE represents such a phase change at constant temperature and it can be noted that line DE is much longer than BC because far more energy is required to change water from the liquid to the gas phase than from the solid to the liquid phase. This is because a greater input of energy is necessary to increase the kinetic energies of individual molecules sufficient to sever the intermolecular bonds of the liquid state (vaporization) in contrast to the energy required to break the bonds in the crystalline state (fusion). After all the liquid has changed into the gaseous state, the temperature of the gas increases—line EF. Exactly the same argument in reverse could be put forward if a hot gas at position F on figure 6.4 is cooled down; instead of line DE representing the latent heat of vaporization, it would correspond to the latent heat of condensation, equal in value to the former, but opposite in sign. This means that heat is released with condensation. Similarly line BC corresponds to the heat of solidification—again heat is released when water changes from liquid to ice.

This example illustrates well the nature of phase change and the associated

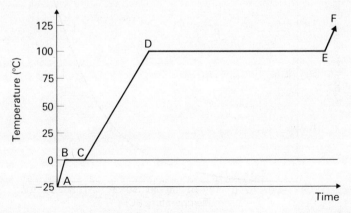

Fig. 6.4 The pattern of temperature and phase changes for water. The temperature remains constant during each phase change as long as pressure remains constant. (after Sears, Zemansky and Young 1974, p. 262)

heats of fusion, vaporization, condensation and solidification. Under ordinary pressure conditions water boils at 100 °C, but in order to explain this, the meaning of vapour pressure has to be considered. Atmospheric pressure is usually taken to mean the total pressure of all the gases in the atmosphere. Thus it is the sum of the partial pressures of the constituent gases—primarily nitrogen and oxygen with small pressures of carbon dioxide, water vapour and other gases. In the case of water vapour in the atmosphere, this partial pressure can vary from zero to a maximum value conditional upon temperature (figure 6.5). If the partial vapour pressure equals the maximum possible for a particular temperature then the air is saturated with water vapour. Figure 6.5 also shows that the equilibrium vapour pressure of 760 mm mercury is associated with a boiling point temperature of 100 °C. In other words, at this stage the equilibrium vapour pressure equals one atmosphere. Thus boiling point can be defined as the temperature at which the equilibrium vapour pressure is the same as the applied pressure. This explains why water boils at lower temperatures in lower pressure conditions which occur, for example, at high altitudes. Another way to view the meaning of boiling point is to consider it as the temperature at which liquid and gas can co-exist; the same applies to freezing or melting points for solid and liquid. Pressure influences the melting point in a similar way to the boiling point. The well known experiment to demonstrate the effect of pressure on melting point involves placing a wire, held down by heavy weights at both ends, over a block of ice. In time the wire passes through the ice but does not cut it in two; the effect of the pressure caused by the wire is to lower the melting point of the ice in the vicinity of the wire to allow its passage through the block. On release of pressure, the water immediately refreezes. The process is known as *regelation*. All substances which contract on

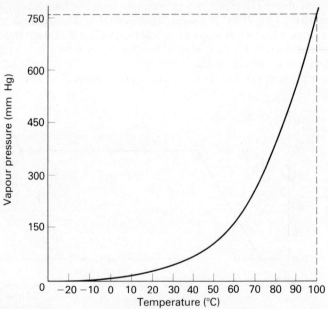

Fig. 6.5 The equilibrium vapour pressure of water with increased temperature. At 100°C the vapour pressure of water is equivalent to 760 mm of mercury (1 atm). (from Anderson, Ford and Kennedy 1973, p. 276)

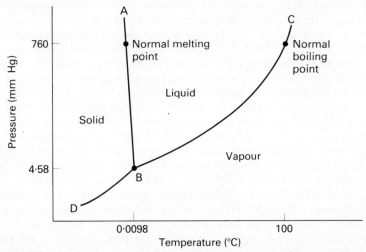

Fig. 6.6 Phase diagram for water; axes are not drawn to scale. (after Anderson, Ford and Kennedy 1973, p. 315)

melting have the property that their melting points are lowered with increase in pressure. With substances which expand on melting, the effect of increases in pressure is to raise their melting points.

Thus both boiling and freezing points vary according to pressure and temperature conditions (figure 6.6). This figure can be described as a simple phase diagram since the conditions under which the various phases exist are indicated. However, such a diagram must be interpreted with care since phase diagrams are usually presented with three axes (pressure, volume and temperature). In the case of figure 6.6, volume is not included and it is assumed that the phases occur in unconfined space. It can be seen that the variation in melting point with pressure is far less marked than the changes in boiling point. Along line AB ice and water exist in equilibrium, but to the left of this line ice is present assuming no volume constraint. Similarly line BC is the equilibrium for the liquid and gaseous states with liquid to the left and vapour to the right. The two lines intersect at point B meaning that at this point the three phases co-exist in equilibrium. This point is known as the *triple point* and with water, the required conditions are a temperature of 0·0098 °C and a pressure of 4·58 mm of mercury. At still lower pressures, according to this diagram, only the solid or vapour states are possible and these are separated by the solid–vapour equilibrium line DB. The process of direct change from solid to gas or vice versa is called *sublimation*. Such a change, like melting or vaporization, requires a heat input called the heat of sublimation which is the amount of heat necessary to change a unit mass from the solid to vapour states without any temperature change. Every pure substance has a characteristic phase diagram; only the case of water has been illustrated. It should also be noted that the phase diagram for pure water will be different from water in which a solute has been dissolved. The effect of dissolving common salt in water is to lower the normal freezing point and to raise the boiling point. These changes result from the decrease in free energy consequent upon the solution of the salt in the water.

The implication of the preceding discussion is that phase changes always take place at melting or boiling points, but this is not necessarily the case. Under certain conditions it is possible to cool liquids below temperatures at which they normally solidify. For example, water can be cooled down as low as − 40 °C without freezing. Such water is termed *supercooled*; it is highly unstable (metastable equilibrium) and any slight disturbance or addition of ice crystals immediately causes the supercooled water to freeze. The freezing point of water can, of course, be lowered below 0 °C by the application of pressure. It is also possible to cool pure saturated air to − 40 °C without condensation taking place and at still lower temperatures the vapour sublimes to ice crystals. The presence of supercooled water is integral to the Bergeron–Findeisen theory which is one theory which accounts for the growth of raindrops under certain conditions. Before the role of supercooled water can be explained in the development of raindrops, further consideration has to be given to the theme of vapour pressure. The relationship between the saturated vapour pressure of water and temperature has been illustrated (figure 6.5), but below 0 °C the nature of the relationship depends on whether it is with respect to a water or ice surface. It is possible to determine the equilibrium vapour pressure of water below 0 °C without the presence of ice. Alternatively, if no pressure is exerted, the equilibrium vapour pressure can be obtained over an ice surface at specific

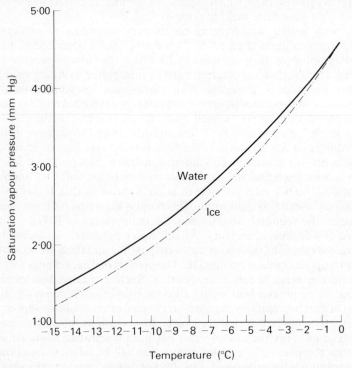

Fig. 6.7 Relationship of saturation vapour pressure with temperature to illustrate the difference between ice and water surfaces. Below 0°C ice crystals can grow through condensation and freezing, but water drops need not so gain. (Data from Weast 1974, pp. 158, 159)

temperatures. These two situations are reflected in the forms of the curves as shown in figure 6.7. The consequence of this is that air which is just saturated with respect to water is supersaturated with respect to ice. Thus when supercooled water droplets co-exist with ice particles in part of a cloud which is vapour saturated, there is a migration of water molecules from the vapour to the ice crystals (sublimation); the result is that the air becomes unsaturated with respect to the supercooled droplets. The response of these droplets is to evaporate thus restoring the saturated equilibrium and producing more water vapour available for sublimation to continue the process of ice accumulation. Eventually these ice crystals will grow to such a size that they begin to fall and when they descend to below the freezing altitude, raindrops result. These drops may further grow by coalescence during descent. Critical to this whole process is the presence of nuclei on to which both condensation and ice crystallization can occur. Crystallization or freezing nuclei appear to be relatively rare in the atmosphere; an important nucleus seems to be the clay mineral kaolinite though other possibilities are meteoric and volcanic dust. Condensation nuclei are far more common and can take the form of smoke, dust, soot, pollen or indeed any particulate matter raised from the earth's surface. It is also common for condensation to take place around particles which exert chemical bonds towards water. Thus the Bergeron–Findeisen theory of raindrop formation depends upon the presence of supercooled water as well as condensation and freezing nuclei, and the process is only possible because of the contrast between the saturated vapour pressure over water and that over ice.

The latter part of this chapter has been devoted to considering the effects of heat transfer, in particular the nature of phase changes. For practical reasons it has been necessary to make sharp distinctions between the three phases though, as the discussion of the Bergeron–Findeisen theory shows, in the physical environment there is nearly always the co-existence of different phases.

6.4 Radiation and the heat balance of soil

In order to exemplify some of the principles raised in this chapter, brief consideration is given to radiation and the heat balance of soil. The rate of receipt of solar radiation at the top of the atmosphere is higher than the rate of receipt at the earth's surface. This latter radiation, known as insolation, is much depleted because of reflection from clouds, absorption in the atmosphere and scattering. Texts such as those by Barry and Chorley (1976), Cole (1970), Monteith (1973) and Sellers (1965) describe these processes in detail. The radiation which arrives at the earth's surface is composed of direct solar radiation (Q), but in addition to this there is diffuse solar radiation (q), radiation which has been reflected as well as absorbed and re-emitted by clouds and particles in the atmosphere. Not all this radiation ($Q+q$) is absorbed by the earth's surface because of the albedo. If the fraction of radiation which is absorbed is represented by a, the quantity of heat which is absorbed is $(Q+q)a$. The albedo equals $(1-a)$. The earth's surface also acts as a radiator and the effective outgoing radiation can be represented by T. These terms have been introduced so that a radiation balance can be given:

$$R = (Q+q)a - T \tag{6.12}$$

where R is the net radiative balance (modified from Sellers 1965). This budgeting approach can be applied on any scale—for the whole world down to a few square centimetres of the earth's surface.

In detail most of the radiation from the earth is in the infrared part of the spectrum. Let this radiation be represented by $I\uparrow$; it is either absorbed by the atmosphere or lost to space. The radiation absorbed by the atmosphere is correspondingly emitted and the recipients are the earth surface again and space. Thus some of the radiation which is emitted from the earth's surface is returned; this phenomenon is known as the 'greenhouse effect' and is extremely important in maintaining the reasonably high surface temperatures. The effective or net outgoing radiation (I) thus results from the difference between that which is actually emitted ($I\uparrow$) and that which is received back from the atmosphere ($I\downarrow$), i.e.

$$I = I\uparrow - I\downarrow \tag{6.13}$$

These principles can be applied to an analysis of the energy balance for part of the earth's surface. The net input of radiation at the earth's surface causes the surface to warm and heat is transferred downwards. The temperature distribution in the soil varies diurnally (from day to day) as well as in a seasonal manner. Some variations in temperature over a few days are shown in figure 6.8. Given these variations on different time scales, it is possible in principle to determine a depth below which such changes do not occur. Clearly such a depth will vary according to locality, but Sellers (1965) suggests that the limit for annual variations ranges from a depth of 5 to 22 m for soil. Visualize a column of soil which extends from the surface to the depth at which isothermal conditions always apply. The heat budget for this volume can be determined.

The net rate (G) at which the heat of this column changes depends upon the following factors:

R, the net radiation balance (from equation (6.12))

L_{E}, the loss due to evaporation at the surface, where L is the latent heat of vaporization

H, the loss due to transfer of sensible heat from the soil surface to the air if the former is warmer than the latter.

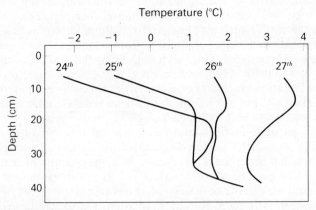

Temperature (°C)

Fig. 6.8 Soil temperature profiles during a period of thaw over four days in February. (after Coutts 1963, p. 47)

If the assumption is made that there is virtually no horizontal transfer of heat from the column to its surroundings, then

$$G = R - L_E - H \qquad (6.14)$$

or

$$R = G + L_E + H \qquad (6.15)$$

Equation (6.15) states that the net available radiative energy warms the soil, evaporates water and warms (or cools) the air above the soil (Sellers 1965). Heat balance equations can be compiled for a variety of environmental situations. For example, during the International Hydrological Decade (1965–1974), the detailed monitoring of drainage basins was encouraged. For basins which contain a glacier which covers at least 30% of the drainage area, the following equation represents the heat balance:

$$F_t = F_r + F_c + F_1 + F_p + F_f \qquad (6.16)$$

where

F_t is the change in heat content due to temperature change in the snow/ice mass
F_r is the radiative heat flux
F_c is the sensible heat flux
F_1 is the latent heat flux from condensation/evaporation
F_p is the heat content of precipitation
F_f is the receipt of heat due to freezing of water

(from UNESCO 1970). Such a heat balance equation can be applied to a whole basin or to any part. In order to work out the equation for a column through a glacier, instruments have to be set up to monitor the necessary variables.

So far the heat budget of the whole unit has been considered. No mention has been made of how heat is redistributed through the unit. In the case of the column of soil, heat is transferred downwards by a variety of processes. There is conduction from one grain to another; infiltration following a shower of rain also rapidly distributes heat through the soil. The movement of soil air helps heat distribution which is further aided by phase changes. Equations have been used to predict the rate of flow of heat which depends on the temperature gradient, volumetric heat capacity of the soil and thermal conductivity. Suppose a soil layer of thickness Δz is considered; over time Δt a temperature change of ΔT occurs. Then the amount of heat which has been lost or gained by the layer is equal to $\Delta T \Delta z C_v$ where C_v is the volumetric heat capacity of the soil. The heat which flows through the layers (Q) can be estimated from:

$$Q = \frac{\Delta T \Delta z C_v}{\Delta t} + \lambda \frac{\partial T}{\partial z} \qquad (6.17)$$

where λ is thermal conductivity of the soil and the term $\lambda \, \partial T/\partial z$ is the heat leaving or entering the layer through its upper or lower surface (from Joynt and Williams 1973). A soil column can thus be subdivided into a large number of layers and by knowing the component values of volumetric heat capacity (see equation (6.4)) and thermal conductivity, heat flow through the column can be predicted.

7
Solids, liquids and gases

The consideration of the physical and chemical properties of matter has necessarily been rather theoretical in approach, though an attempt has been made to illustrate the environmental significance wherever possible. This strategy is combined in order to explain the nature of solids, liquids and gases. In the physical environment, these phases nearly always occur in particular associations; however practical considerations necessitate that these are tackled in turn. An appropriate way to proceed is to summarize basic theoretical concepts for each phase concerning the nature and properties of matter at the atomic or molecular scale; this leads to consideration of the properties at the macroscopic scale.

7.1 Solids

Liquids and gases are characterized by their fluidity, a property not shared with solids which are distinguished by their structural order. The conditions under which solids occur can be understood with reference to phase diagrams (figure 6.6). Unlike gases, individual solids have characteristic volumes and are nearly always incompressible, properties which result from the strong interatomic or intermolecular forces acting in solids. In the physical environment, individual solids are best exemplified by hard rocks such as granites or gneisses, though these rocks are made up of particular mineral assemblages. If supercooling does not occur, a solid is formed at and below the freezing point which is dependent also upon pressure conditions. The phase change from liquid to solid involves the release of heat of fusion. This implies a lower energy status for solids in contrast to their liquid phases. Gases and liquids possess the ability to diffuse rapidly, but the diffusion of solids is extremely slow and is rarely perceptible.

The distinctive feature of solids is their structural order reflected in their crystallinity. A crystal is a specific three-dimensional pattern of particular atoms, molecules or ions and the structure is constant for particular solids. The planar surfaces of crystals are known as faces and these intersect at angles characteristic of the substances. Crystals, when broken, split along cleavage lines which, like faces, are characteristic of specific crystals. An understanding of crystal structure is essential to any consideration of solid matter, but before this is tackled, it should be noted that there are also non-crystalline solids and these are described as being *amorphous*. Glass, for example, is amorphous since the molecules can assume an infinite number of dispositions within the material. A chemical approach is to view amorphous solids as highly viscous liquids rather than solids (Anderson, Ford and Kennedy 1973), but this point is of little importance to the present discussion since most solid matter in nature is crystalline.

7.1.1 Crystal structure

In chapter 2 it was noted that the composition of the earth's crust is dominated by a small number of elements; in fact oxygen, silicon, aluminium and iron make up 87% of the mass of the crust. However, such apparent simplicity is misleading since these elements as well as others are present in a large number of combinations in the form of *minerals* which are homogeneous naturally occurring crystalline solids. Mineralogy forms a major subject within geology and soil science and in accord with the approach throughout this book, it is not the aim to summarize the properties of specific minerals, but rather to focus attention on the overall nature of crystals.

The geometrical order in a crystal is often described as a lattice and investigation into such structures is possible by the use of X-ray diffraction techniques. In particular the three-dimensional disposition of atoms or molecules within a crystal is known as the *space lattice*; the specific points where the atoms or molecules occur are called lattice sites. Within a space lattice, the smallest recurring pattern is called the *unit cell*; it is the linkage of these unit cells which produces the space lattice. The form of unit cells conditions the geometry of the whole lattice. For example, figure 7.1 illustrates the unit cell for sodium chloride and a macroscopic crystal of sodium chloride can have the same cubic shape. However, it is important to note that a mineral does not necessarily have one crystal form. The sodium chloride crystal need not occur in the same geometrical proportions as the unit cell. In fact, sodium chloride can occur in the three forms shown in figure 7.1. Small cubes result (figure 7.1 (a)) when crystal growth is equally possible in all three directions; the result illustrated in figure 7.1 (b) occurs when crystals grow in a shallow dish containing sodium chloride solution because their growth is impeded in one direction; octahedral crystals (figure 7.1 (c)) are formed when they are grown suspended in a solution which contains some urea. Because sodium chloride has these various crystalline forms, it is described as *polymorphic*. Despite this complication, it is useful to identify various types of unit cell symmetry.

Figure 7.2 (a) illustrates simple cubic symmetry. The three axes are

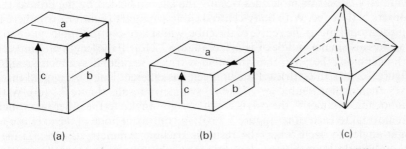

(a)   (b)   (c)

Fig. 7.1 Various shapes exhibited by sodium chloride crystals grown under different conditions. The crystal axes are given by *a*, *b* and *c*. (from Anderson, Ford and Kennedy 1973, p. 374)

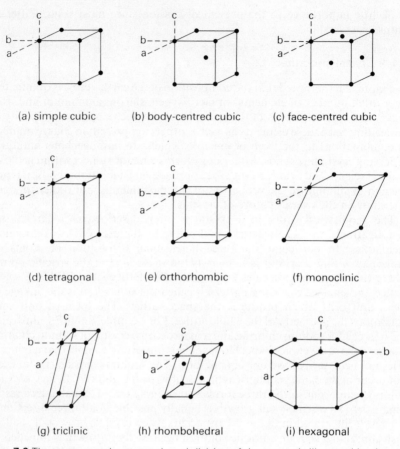

(a) simple cubic   (b) body-centred cubic   (c) face-centred cubic

(d) tetragonal   (e) orthorhombic   (f) monoclinic

(g) triclinic   (h) rhombohedral   (i) hexagonal

Fig. 7.2 The seven crystal systems: the subdivision of the system is illustrated by the cubic unit cells (a), (b) and (c).
(a) simple cubic, (b) body-centred cubic, (c) face-centred cubic, (d) tetragonal, (e) orthorhombic, (f) monoclinic, (g) triclinic, (h) rhombohedral, (i) hexagonal.

indicated by *a*, *b* and *c*—it is along these axes that the unit cell would be repeated many times to build up the complete lattice. With simple cubic symmetry, atoms or molecules occupy the sites indicated by the dots at the corners of the cube. With body-centred cubic symmetry (figure 7.2 (b)) there is an additional site at the centre of the cube, whilst face-centred cubic symmetry has sites also in the middle of each face (figure 7.2 (c)). If the length of a unit cell is longer along the *c*-axis than the other two axes, tetragonal symmetry results (figure 7.2 (d)). In contrast if the unit cell has different lengths along all three axes, then orthorhombic (or rhombic) symmetry results (figure 7.2 (e)). With monoclinic symmetry, the *c*-axis is not at right-angles to the other two which are normal to each other (figure 7.2 (f)); in contrast if none of the axes are at right-angles to each other, the result is triclinic symmetry (figure 7.2 (g)). Rhombohedral symmetry is identical to the cubic type, except that the axes are all equally inclined to each other, but not at 90° (figure 7.2 (h)). The final type is hexagonal symmetry in which cell lengths along the *a*- and *b*-axes are the same, but are not equal to the length along the *c*-axis; further, the angle between the *a*- and *b*-axes is 120° whilst between *b* and *c*, and *c* and *a* the angle is 90° (figure 7.2 (i)). The variation in lattice sites within particular symmetries has been illustrated for the cubic crystal system; variations are also possible with the other basic crystal symmetries. Some examples of crystal form are given in table 7.1.

The approach to crystals so far has been to view them as lattices with atoms or molecules occurring as dots at the lattice sites. Though useful models, they are also misleading since atoms and molecules occupy space, and thus it is important to appreciate that the nature of packing is significant with respect to crystal structure. The well used example of the sodium chloride crystal is composed of ions which are closely packed—one way to view this crystal is to consider it made up of an array of chloride ions with the sodium ions occurring in the spaces between the chloride layers (Sienko and Plane 1974). In detail the nature of the packing is also dependent upon the size of the ions; anions and cations of the same size will pack in a different way to the arrangement that occurs if the anions are considerably larger than the cations. The sizes of ions can be expressed in terms of their radii, and the ratio of ionic radii is a useful guide to spatial arrangements, the details of which can be followed up in Gymer (1973, pp. 448–50).

The form of this packing is important not only with regard to chemical properties, but also with respect to physical properties; for example the strength of a material depends upon the degree of atomic or molecular interlocking. Another point worthy of mention without any elaboration is that it is not necessary for every lattice site to be occupied by an atom or molecule. The presence of lattice vacancies in a crystal is a criterion for stating that the crystal has a *defect* to indicate the difference between a perfect and an actual crystal. Defects also arise when atoms are forced to occupy positions between lattice sites (interstitial positions). Other forms of defects are possible, in fact all crystals have defects to varying degrees; for example, a sodium chloride crystal would always have some ions missing from lattice sites. An impurity in the crystal could occupy some of the interstitial positions.

For long the focus of the study of crystals (crystallography) was their description and classification, and consideration has been given to the basic crystal symmetries. However, in order to understand the nature and properties of crystals, and indeed of substances in any state or form, the nature of the bonds between atoms or molecules needs to be examined. This immediately

Table 7.1 Examples of minerals with different crystal forms. Sources: Smith (1971) and Pierce (1970).

Symmetry	Minerals
cubic	magnetite, fluorite
tetragonal	rutile, cassiterite
orthorhombic	limonite, topaz
monoclinic	sphene, actinolite
triclinic	kyanite, microcline
rhombohedral	dolomite, siderite
hexagonal	apatite, graphite

allows an identification of the four types of crystals according to bonds, viz. ionic, molecular, covalent and metallic crystals.

Ionic crystals

The sodium chloride crystal is an example of an ionic crystal since it is made up of anions and cations. Within this crystal, there are equal numbers of anions and cations and their opposite charges lead to a strong electrostatic force between ions. When the electrical charges on the ions are not equal, the lattice must adopt a structure so that there is electrostatic equilibrium. The strength of the electrostatic bond in ionic crystals results in these substances being distinguished by their moderately hard but brittle nature.

Molecular crystals

With molecular crystals, molecules occupy the lattice sites and the bonds between these result from dipole–dipole interaction, hydrogen bonding or from van der Waals' forces; these forms of attraction require some explanation. A dipolar molecule is one which has two separate centres of electrical charge of equal and opposite values. Water is a good example to illustrate the nature of polarity. Water is a triatomic molecule and the angle of the bond between the H—O—H atoms is about 105° (figure 7.3). The effect is to give a water molecule opposite partial charges at either end. With ionic bonding there is complete transfer of electrons from one atom to another, but in many cases such total electron transfer is not possible—instead bonds exist through the sharing of electrons. With hydrogen gas (H_2), individual molecules are *covalently bonded* through the sharing of an electron pair. Covalent bonds link the two hydrogen atoms to an oxygen atom for each water molecule. As already noted intermolecular forces will arise in water because of dipole–dipole interaction: the negative end of the molecule (oxygen atom) will be attracted to the positive end (hydrogen atoms) of another molecule. The result is that the hydrogen atoms are attracted to the oxygen atoms—in particular the hydrogen atoms are attracted to the unshared electron pairs of the oxygen to form *hydrogen bonds*. This particular form of attraction between

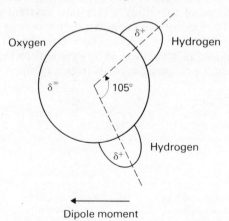

Fig. 7.3 The water molecule is made up of one oxygen atom and two hydrogen atoms: the polar nature of the molecule is explained by the partial negative charge ($\delta^=$) on the oxygen and the partial positive charge (δ^+) on the hydrogen atoms.

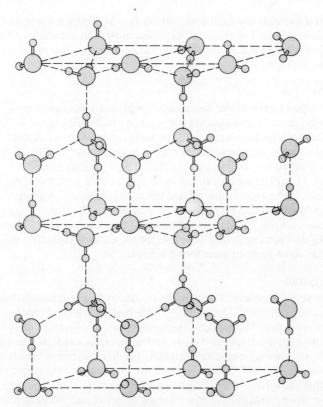

Fig. 7.4 A small part of a crystal of ice. The molecules are shown with approximately their correct sizes (relative to the interatomic distances). Note hydrogen bonds, and the open structure that gives ice its low density. The molecules are indicated diagrammatically as small spheres for oxygen atoms and still smaller spheres for hydrogen atoms. (from Pauling 1970, p. 431)

molecules results when hydrogen is covalently bonded to a markedly electronegative atom such as oxygen. By electronegativity is meant the readiness with which an atom can accept an additional electron to form a negative ion (Harrison 1972). Electron movement in one atom can be so synchronized with the rotation of an electron in an adjacent atom that at one particular instant the negative end of one atom is attracted to the positive end of the other. At the next instant the charge distributions are reversed, but the effect is still an attraction; this is an example of a van der Waals' force. Such a force is weak compared to the other forms of bonding. A return can now be made to an example of a molecular solid, ice, to illustrate the significance of some of these principles.

Ice is distinguished by its regular crystal structure. In the liquid state only a few of the possible hydrogen bonds operate and these are always changing. In contrast, in the solid state, all possible hydrogen bonds are activated to give the crystal its distinctive rigid form (figure 7.4). In particular each oxygen atom is enclosed tetrahedrally by four other oxygen atoms. The open nature of the structure demands more volume than the disordered liquid structure so that water expands on freezing and also ice is less dense than water.

Molecular crystals are held together by bonds which are not as strong as other types of bonds. The result is that these molecular crystals tend to be soft and have low melting points, the latter because little energy is required to break the bonds.

Covalent crystals

With this type of crystal the lattice sites, which are occupied by atoms, are covalently bonded. In effect the crystal is a huge molecule since it is made up of one colossal, interlocking structure. The usual example of a covalent crystal is a diamond which is an allotrope of carbon. In this crystal each carbon atom is covalently bonded by sharing pairs of electrons with four other carbon atoms (figure 7.5). The robustness of the structure combined with the strength of the bonds results in the exceptionally hard nature of diamond. A similar structure is exhibited by quartz (SiO_2) since a silicon atom, which occurs in the middle of each tetrahedron, is bonded covalently to the four oxygen atoms at the corners. Again such a structure accounts for the nature of quartz—its chemical stability has already been mentioned (chapter 5).

Metallic crystals

The lattice sites of metallic crystals are occupied by cations, which are good conductors of electricity. The structure of a metal is thus a lattice of metallic cations around which there exists a swarm of very mobile electrons which give the metal its high electrical as well as thermal conductivity. In the solid state the metallic ions lose at least one electron from their outermost shells, resulting in the positively charged ions as well as the cloud of negative electrons. The theory is that the forces of attraction between the cations and the electrons are sufficiently strong relative to other forces of repulsion between the cations themselves and between the electrons themselves to cause the cohesion of the solid (Harrison 1972). Thus the bonding of a sodium crystal can be viewed as

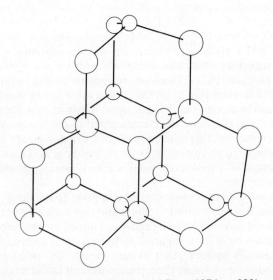

Fig. 7.5 Structure of diamond (from Sienko and Plane 1974, p. 209)

resulting from the attraction between the electrons and the sodium cations. Metals are very variable in terms of their hardness, but certain metals are, of course, distinguished by their strength. Iron owes its particular strong nature to metallic bonding, but in detail the nature of the constituent crystals must also be taken into account. For example, metallic bonding is weaker along crystal boundaries and thus iron which is imperfect in crystal structure because of variations in packing arrangements, will possess lower strength than a more pure iron which has a near-perfect crystal structure. To further increase the strength of iron, carbon and other metallic elements can be introduced to improve crystalline interlinkage and bonding.

The aim of outlining crystal types according to the nature of their bonds was to illustrate the theme that the properties of crystalline solids are dependent upon their atomic configurations and bonding. The relevance of this to the physical environment, for example, is that it helps the understanding of weathering processes, the energetics of which have been described (chapter 5). Weathering of minerals results in the breaking of the constituent bonds and thus to delve into the topic in greater detail requires some assessment of crystal or lattice energy. This can only be touched upon, but the principles are also of direct relevance to phase change.

The suggestion has already been made that the bonds are of variable strength and there is also variability according to mineral type. Ice again provides a useful example to illustrate the applicability of these concepts to phase change. As described, ice is a highly ordered structure with molecules of water linked chiefly by hydrogen bonding. A change of state from solid to liquid demands that sufficient energy is applied to break enough bonds to produce the disordered liquid structure. The energy required to break these bonds, as will be recalled, is the latent heat of fusion which equals $333 \cdot 7 \text{ J g}^{-1}$. One practice which will be encountered is to express the energy of a crystal in terms of the energy required to convert one mole of the substance from the solid to the gaseous state. Another closely related approach to express crystal stability is to predict values for *bond strengths* since these measures will clearly indicate the magnitude of attraction between atoms or molecules. However, it is essential to recognize the type of bond to which the measure refers. In the case of ice, crystal energy can be described in terms of the amount of energy needed to break intermolecular bonds; a change of state is involved and there is no chemical alteration. When dealing with chemical weathering, changes in the nature of molecules result and thus the strengths of bonds between atoms assume significance. As described in chapter 5 values for free energy of formation are indicative of mineral stability which, of course, depends upon bond strengths.

Stress has been placed so far on the importance of the atomic and molecular structure of solids. Before attention is focused on the physical properties of solids, it is instructive to illustrate some of these structural considerations with respect to some examples: clays are a good choice since their nature and behaviour are intimately connected with their atomic structures. However, it should be noted that clays cannot be considered as pure solids since they are compressible and under certain situations possess fluidity.

Clays are distinguished by being very small particles, smaller than 0·002 mm in equivalent size diameter, and are also distinguished by being flake-like in

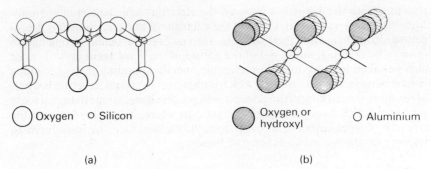

Fig. 7.6 The two basic structural units of silicate clays.
(a) silica tetrahedron: a silicon atom surrounded by four oxygen atoms.
(b) alumina octahedron: an aluminium atom surrounded by six hydroxyls or oxygens.
(after Brindley and MacEwan 1953, pp. 17, 18)

shape. It is because of their flake-like shape that the size of clay particles is described in terms of equivalent size diameter since the determination of clay content using sedimentation techniques has to assume that the particles can be considered as spheres of particular diameters. In detail the form of these flakes varies according to the type of clay, but nevertheless, the outstanding consequence is that clays present a huge surface area. As will be shown, this is of particular importance since nutrient storage takes place at the surface of the clay. Brown (1974) states that the total surface area of 1 g of clay is commonly in the region of 50–100 m^2 though higher values are possible.

Clays can be subdivided into two broad groups—the silicate clays which occur in temperate areas and iron and aluminium hydrous oxide clays which are best developed in tropical and subtropical areas. For this discussion consideration is limited to silicate clays. These are crystalline and it is instructive to outline the nature of these crystal structures. Most silicate clays are in fact alumino-silicates because aluminium and silicon are present. These are associated with two basic structures (Brady 1974). A tetrahedron results when one silicon atom is surrounded by four oxygen atoms (figure 7.6 (a)) whilst an aluminium octahedron has an aluminium atom in the centre and is surrounded by six oxygens or hydroxyls (figure 7.6 (b)). These structures are the basic building blocks for many clays; for example a layer can be formed by interlocking silica tetrahedra which are held together by the sharing of oxygen atoms. A similar plan or form results when many alumina octahedra are linked. These two types of layers occur in various combinations providing a very useful classification of silicate clays. Four groups are identified according to pattern of alternation of tetrahedral and octahedral layers.

1 : 1 type silicate clays
These are described as 1:1 because they are made up of the alternation of one silica tetrahedral layer with an alumina octahedral layer. The best known type of clay in this group is kaolinite which is made up of two sheets of the basic molecular components held together by the sharing of oxygen atoms; individual two-layer units are attracted to others by hydrogen bonding between oxygen and hydroxyl ions in adjacent double sheets (figure 7.7).

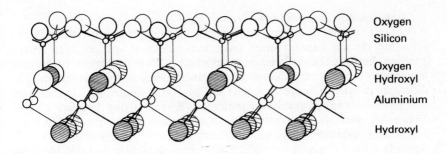

Oxygen
Silicon

Oxygen
Hydroxyl

Aluminium

Hydroxyl

○ Oxygen

◍ Hydroxyl

○ Aluminium or other ion
  in 6-coordination

○ Silicon or aluminium in
  4-coordination

A key to the representation of atoms in figures 7.7 - 7.10.
Note that density of shading and outline decreases with
the depth of the atom represented

Fig. 7.7 Three-dimensional view of kaolinite; the lines joining atoms represent bonds
though not all are shown. (after Brindley and MacEwan 1953, p. 19)

2:1 type expanding silicate clays
This type has a sandwich structure whereby an alumina sheet occurs between
two silica layers, and a good example is montmorillonite (figure 7.8). The
bonding between the sandwiches is different to kaolinite in that the crystal
units in montmorillonite are weakly linked by O—O bonds; no hydrogen
bonds are present. This weak bonding means that the units can be easily
separated. This group of clays is distinguished by being able to attract water
molecules and cations between the crystal units resulting in the expansion of
the lattice.

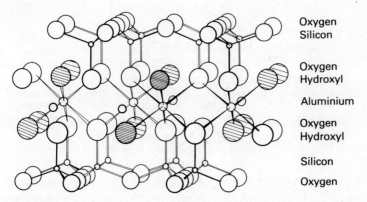

Oxygen
Silicon

Oxygen
Hydroxyl

Aluminium

Oxygen
Hydroxyl

Silicon

Oxygen

Fig. 7.8 Three-dimensional view of montmorillonite (modified from Brindley and
MacEwan 1953, p. 23)

E

2:1 type non-expanding silicate clays

As can be imagined these clays have exactly the same structure as the previous type, but do not expand since water molecules are not attracted to the positions between the sandwiches. An example is the mineral illite; in this case aluminium atoms occupy some of the silicon sites in the tetrahedra. The difference in valency between silicon (Si^{4+}) and aluminium (Al^{3+}) means that this partial exchange results in tetrahedra with a strong negative charge. Potassium ions are then attracted to convenient sites thus bonding the sandwiches together. The transformation of montmorillonite into illite depends upon the incorporation of potassium; such a process takes place, for example, on the sea floor so that potassium from river discharge does not continually increase the amount of potassium in the sea. Illite is far more stable volumetrically than montmorillonite, though a limited amount of expansion is possible. Illite is an example of a mica clay mineral, and the term is also used to describe the group of such minerals. The structure of the mica clay mineral, muscovite, is illustrated in figure 7.9 and the similarity between montmorillonite and this mineral is immediately apparent.

2:2 type silicate clays

This group can also be described as having a 2:1:1 layered structure; an example is chlorite which can be viewed as consisting of alternating 2:1 layers with octahedral layers (figure 7.10). In the octahedral layers Al^{3+} in part take the place of Si^{4+} to give a negative charge. The other layers often include Mg^{2+} and Fe^{2+} as well as Al^{3+} to result in a positive charge. The oppositely charged

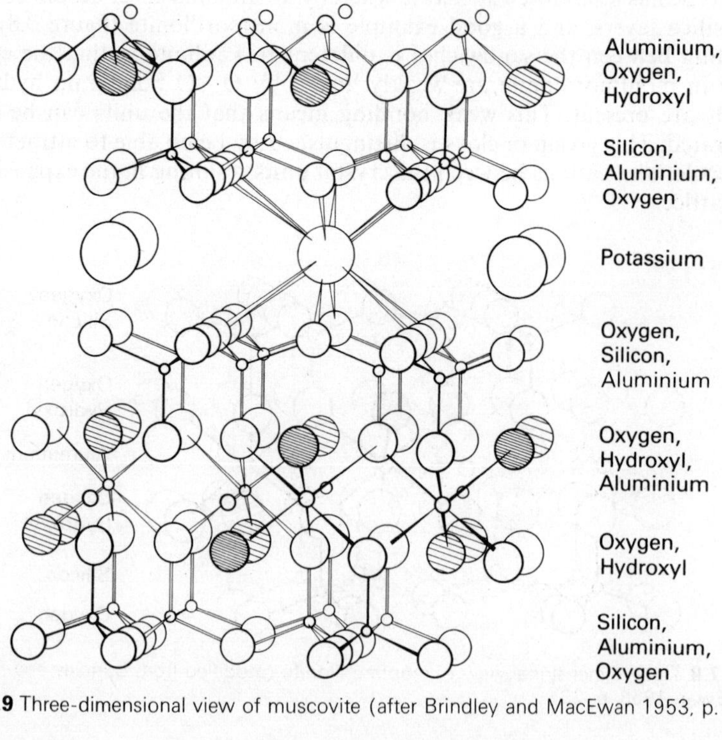

Fig. 7.9 Three-dimensional view of muscovite (after Brindley and MacEwan 1953, p. 23)

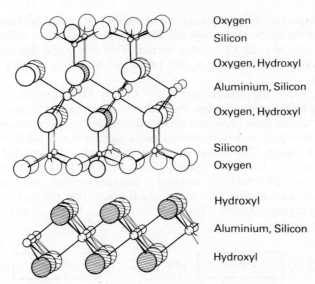

Oxygen

Silicon

Oxygen, Hydroxyl

Aluminium, Silicon

Oxygen, Hydroxyl

Silicon

Oxygen

Hydroxyl

Aluminium, Silicon

Hydroxyl

Fig. 7.10 Three-dimensional view of chlorite (after Brindley and MacEwan 1953, p. 24)

layers lead to strong bonds which are further assisted by hydrogen bonding (Birkeland 1974), so that the mineral is volumetrically stable. The concentration of magnesium in the sea is regulated by its incorporation into chlorite (Gymer 1973).

Before some of these structural considerations are interpreted in terms of specific properties of clays, it is important to note that there are many other clay minerals which have not been mentioned; indeed the study of clays is a major subject in itself. Also it is relevant to stress that in actual soil, clay types occur in intimate association; it is possible to identify mixed-layer minerals since individual crystal units can be made up of more than one mineral type. A large number of cation substitutions are possible leading to a variety of types of clay.

Individual clay particles are called micelles and usually possess a negative charge; thus cations as well as water molecules are attracted to micelles. The term *adsorption* describes the attraction of cations and water molecules to clays. In temperate areas cations such as H^+, Al^{3+} and Ca^{2+} tend to occur most frequently followed by Mg^{2+} and then K^+ and Na^+. Plants obtain nutrients by cation exchange between their roots and micelles and this process is of fundamental significance. Brady (1974) neatly illustrates the principles involved by considering the fate of one calcium ion adsorbed on a micelle when there are hydrogen ions around; the result is the replacement of the exchangeable Ca^{2+} by the H^+ ions (figure 7.11). This simple example also illustrates the effect of pH on nutrient availability—when the pH is low (high concentration of H^+ ions), many nutrients adsorbed in micelles will be lost through leaching, i.e. they are. exchanged for H^+ ions. The ability to participate in cation exchange between micelles and the soil solution is called the *cation exchange capacity*: this is the sum of all exchangeable cations that a micelle can adsorb. As explained this is dependent upon pH conditions, but more importantly upon the nature of the clay. It should be noted in passing that the cation exchange capacity of soil is also very much influenced by the nature and amount of organic matter. This brings the discussion back to considering the structure of clay types and how they possess net negative charges.

The effect of the replacement of Si^{4+} ions by Al^{3+} within the structures has already been mentioned; the same effect results when Al^{3+} ions are replaced by Mg^{2+}, and these processes are called *isomorphous substitution*. Clearly the degree to which such substitution takes place will control the ultimate net negative charge. This charge can be further increased if unsatisfied valancies are exposed at the broken edges of layers. With 1:1 type silicate minerals, oxygens and hydroxyls are exposed at external surfaces; the hydrogen of the hydroxyls can, in part, dissociate, which means that the surfaces of micelles develop negative charges. This process is of particular relevance to 1:1 minerals such as kaolinite.

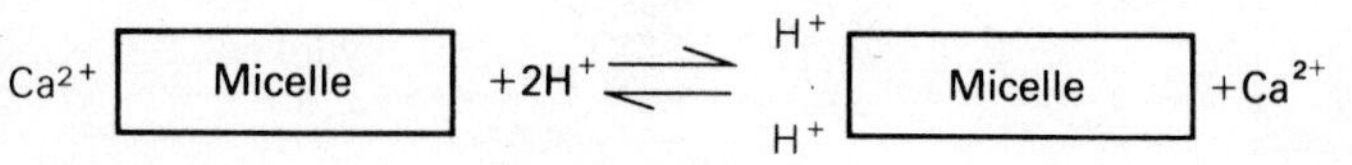

$$Ca^{2+} \boxed{\quad \text{Micelle} \quad} + 2H^{+} \rightleftharpoons \begin{matrix} H^{+} \\ \\ H^{+} \end{matrix} \boxed{\quad \text{Micelle} \quad} + Ca^{2+}$$

Fig. 7.11 Cation exchange on a micelle of one calcium for two hydrogens (from Brady 1974, p. 97)

As already described, kaolinite is distinguished by its rigid and fixed lattice structure. Cations cannot penetrate its structure and thus cation exchange is limited to the perimeter of the clay particles. This, coupled with the fact that kaolinite particles are large in contrast to other clays, means that kaolinite has a low cation exchange capacity and also that its swelling and shrinking properties are low. In marked contrast, much smaller particles of montmorillonite, besides having a high cation exchange capacity (of the order of 10–15 times that of kaolinite) also are highly plastic, cohesive and change in volume according to amount of water present (Brady 1974). From the agricultural standpoint, soils containing significant quantities of montmorillonite are potentially very fertile, but require very careful management. Similar structural considerations of illite and chlorite would lead to the broad conclusion that these minerals are similar in cation exchange capacity to montmorillonite; indeed their overall properties can be viewed as having values between the extremes for kaolinite and montmorillonite.

7.1.2 The strength of solids

Consideration of clays at the atomic level aids understanding of some of the macro-physical properties such as volume stability. To conclude this part of the chapter dealing with solids an important property of solids, namely their strength is selected; in detail strain–stress–strength interrelationships are examined. A knowledge of these principles is essential for an analysis of slope processes and stability. The application to geomorphology of these engineering principles is well demonstrated by Carson (1971) and at a more introductory level by Whalley (1976).

In chapter 3 a brief introduction was given to normal and shearing stresses, but before these principles are developed some attention has to be given to the behaviour of solids when stress is applied. When a solid is subjected to sufficient stress, it will begin to deform in response to such stress. The term *strain* describes this relative change in body dimension or shape. If the stress is compressional, then the amount of strain is defined as:

$$\varepsilon = \frac{\Delta l}{l_o} \qquad (7.1)$$

where ε is linear strain, Δl is change in length and l_o is initial length. Clearly ε is dimensionless. Equation (7.1) expresses strain as a ratio of change in length in one direction. It is more realistic to resolve strain, like stress, into three principal components often labelled ε_1, ε_2 and ε_3. For many problems it is important to know not only the magnitude of ε but its rate of change with time ($d\varepsilon/dt$), conventionally denoted by $\dot{\varepsilon}$.

An appropriate method to introduce basic concepts of stress–strain–strength interrelationships is to begin with a simple example. Suppose that compacted, clean, dry sand is placed in a specially constructed box as illustrated in figure 7.12. As can be seen this box is designed to permit a fracture through the sand. This fracture will result from the application of a force on the lower half of the box whilst a restraining force is applied to the upper half. This restraining force is measured on a proving ring, a force-measuring device in which a dial indicates the deformation of a steel circle. A plate is placed over the sample and various loads can be applied to this plate. For any one normal load the force from the piston can be gradually increased until there is failure of the sample; at this point the *peak shear load* can be recorded from the reading on the proving ring. After the point of failure has been passed, the resistance to shearing drops to a constant value, when the *ultimate* or *residual shear load* is read from the proving ring. The distinction between these two situations is illustrated in figure 7.13. This type of experiment can be further demonstrated by using actual figures as given in an example by Capper, Cassie and Geddes (1971): suppose the shear box is 0.0645 m^2 in plan and experiments are carried out with three normal loads to give the following results:

Normal load (kg)	500	1000	1500
Peak shear force (kN)	4.92	9.80	14.62
Ultimate shear force (kN)	3.04	6.23	9.36

These values have to be converted into stresses: for example the normal stress resulting from a mass of 500 kg acting on an area of 0.0645 m^2 is $(500 \times 9.81)/0.0645 \text{ N m}^{-2}$ or 76.0 kN m^{-2}. The shear forces have to be divided by the area to obtain the shear stress values which are given below:

Normal stress (kN m^{-2})	76.0	152.1	228.1
Peak shear stress (kN m^{-2})	76.3	151.9	226.7
Ultimate shear stress (kN m^{-2})	47.1	96.6	145.1

These values can then be plotted on a diagram (figure 7.14). The angles of inclination of these lines indicate the ratios of shear stress to normal stress—

E*

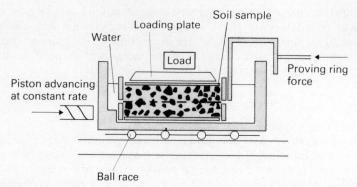

Fig. 7.12 The direct-shear apparatus. For the experiment with sand as described in the text, the water is not present. (from Carson and Kirkby 1972, p. 70)

such values are *angles of shearing resistance* conventionally denoted by ϕ and also called the *angle of internal friction*. In this example two values of ϕ are obtained corresponding to peak shear conditions ($\phi_1 = 45°$) and ultimate shear conditions ($\phi_2 = 32°$). From this example it is clear that the resistance to shear of materials will increase according to the normal stress. The strength of a material will thus depend on the angle of shearing resistance which will be conditioned by such factors as degree of compaction, shape and size of individual particles, amount of particle interlocking, friction between particles and normal stress. The differences between the peak and ultimate values can in part be explained by the analogy between the coefficients of static and dynamic friction. More important is the re-orientation of particles along the shear plane once movement has been initiated (Carson 1971).

Another influence on the strength of a material is its *cohesion* which depends upon the forces of interparticle attraction and is essentially independent of normal load. The sand in the shear box possessed no cohesion and thus the

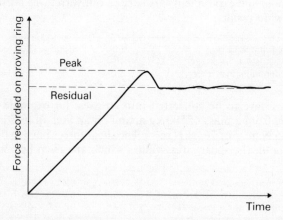

Fig. 7.13 A graph showing the increase in shear force applied to dry sand to cause failure. The distinction between the peak and residual (or ultimate) shear forces is illustrated.

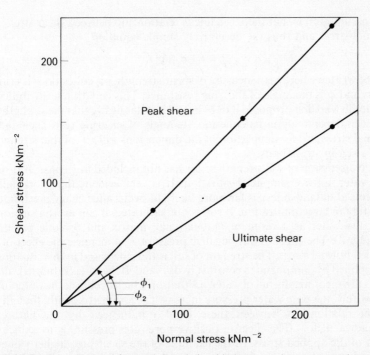

Fig. 7.14 Graphs to illustrate for sand the relationship between shear stress and normal stress using the experimental data. (after Capper, Cassie and Geddes 1971, p. 60)

graphs of figure 7.14 pass through the origin. In contrast, damp sand possesses cohesion, and this is the reason, of course, why sand castles can be built from damp sand but not from dry sand. Nearly all naturally occurring material possesses cohesion and if the experiment as described above is repeated with such samples, then the graphs will intersect the shear stress axis (figure 7.15).

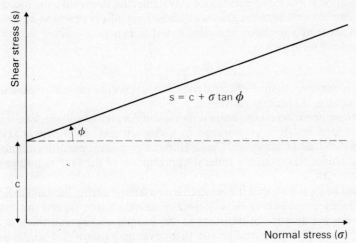

Fig. 7.15 Graphical representation of equation (7.2).

The assumption is that there is a linear relationship between shear stress and normal stress and thus the deceptively simple equation

$$s = c + \sigma \tan \phi \tag{7.2}$$

can be written where s is a measure of shear strength, c is cohesion, σ is normal stress and ϕ is the angle of shearing resistance. This equation is attributed to Coulomb who first proposed it in 1776 and it states neatly that the strength of a material depends upon its cohesion, its angle of shearing resistance and its normal stress; brief examination of the dimensions will reveal that strength is also given in stress units.

The discussion of soil strength so far has not included any consideration of soil water, an oversimplified situation. It is very evident that slope failure (expressed in landslides), is far more likely to occur after heavy precipitation resulting in large infiltration. If soil is not saturated, it can exert a suction to hold the water as a result of gravitational, matric and osmotic potentials (chapter 4). The effect of this negative pressure is to increase the effect of the applied normal stress. The strength of soil is thus increased in this situation—exemplified by damp sand in contrast to dry sand. If the soil is saturated, all the pores are theoretically full of water, a situation which occurs at the water table. Below this, the pore water pressure increases according to depth, though the actual relationship between these is also influenced by the nature of groundwater flow. The effect of positive porewater pressure is to reduce the effect of the applied stress. In the example of the shear box, higher values of applied normal stress necessitated higher shear stresses in order for fracture to occur. The applied normal stress thus controls the resistance to failure through its effect in forcing particles to interlock. The presence of porewater pressure is to reduce the effect of applied normal stress and thus it is more useful to define the parameter σ' as the *effective* normal stress, to distinguish it from σ, strictly the *total* normal stress. Clearly σ' is less than σ, and for sands the difference is equal to the porewater pressure (u), i.e.

$$\sigma - \sigma' = u \tag{7.3}$$

This situation is more complicated for clays. Effective stress has a distinct effect on the size of voids between grains which in turn affects cohesion and hence angle of shearing resistance. It is thus useful to express equation (7.2) in the more general form

$$s = c' + \sigma' \tan \phi' \tag{7.4}$$

where s is shear strength, c' effective cohesion, σ' effective normal stress and ϕ' effective angle of shearing resistance.

The topic opens up a broad and difficult subject area of soil mechanics and the interested reader is encouraged to follow up texts by Terzaghi (1943), Jumikis (1967) and Capper and Cassie (1963). For present purposes it is hoped that the reader has gained a general appreciation of the factors influencing shear strength.

It must be clear now that the results derived from shearing the sand in figure 7.12 are very unrealistic since soils usually contain water. Indeed the critical factors are the amount of moisture in the sample undergoing the test and whether drainage is permitted or not. If, for example, a saturated sand is being tested, but the sample is sealed to prevent any drainage, then there will be no increase in strength with increases in normal stress because any increment in the latter will be counteracted by an increase in porewater pressure. Such a laboratory situation is not a good model of field conditions since the increase in porewater pressure would lead to water movement. This illustrates one of the limitations of the direct shear apparatus; more sophisticated equipment is used, for example the triaxial apparatus which permits better experimental control (Whalley 1976), but given these problems about sample conditions, it may be that soil strength is best assessed in the field, at least for geomorphological purposes.

The strength of a material is the resistance which it can offer to deformation and slippage. The former is expressed as strain or compressive strength whilst the latter is shear strength. Material, of course, varies in its response to the application of stresses. Elastic behaviour implies that deformation is fully recoverable up to a certain limit—in other words strain increases in a linear manner according to increasing stress. With viscous materials which are fluids, the application of stress leads to continuous strain. In other words there is no upper limit to deformation resulting from the application of stress; in this case the rate of strain ($\dot{\varepsilon}$) is a function of stress. Another ideal type of material is described as being rigid plastic; here no deformation occurs until a particular stress is achieved. If it is impossible to apply a stress higher than this critical value, then the material is plastic. These ideal material types are described in greater detail by Carson (1971); in practice, of course, materials combine these qualities so that terms such as visco-plastic, elasto-plastic or elasto-viscous are encountered.

The application of engineering principles can be demonstrated with reference to hillslopes. For example, in south Wales the landscape is characterized by deep valleys with steep valley-side slopes of distinctive uniformity. Rouse and Farhan (1976) have investigated a small area in south Wales in order to establish if the uniformity of valley-side slopes can be interpreted in terms of minimum values at which slope failure occurs. At a large number of sites, samples were extracted so that measures for bulk density, moisture content and residual angle of shearing resistance were obtained. The saturated bulk density (γ_s) was obtained from

$$\gamma_s = \gamma_d + \gamma_w - (\gamma_d/2\cdot65) \tag{7.5}$$

where γ_d is dry bulk density and γ_w is the density of water. They obtained saturated bulk densities (γ_s) ranging from 15·2 kN m^{-3} to 19·3 kN m^{-3}; the values for the residual angle of shearing resistance (ϕ_r') varied from 31° to 40°, a range to be expected given the granular nature of the material which also had a low clay content.

The next step in the analysis by Rouse and Farhan (1976) was the selection of a suitable model so that they could predict the limiting angle of slope—the minimum angle at which slope movement could occur. Clearly assumptions had to be made about the nature of slope failure and they considered that such movement was most likely to be almost planar and parallel to the ground surface given the straight rock hillslopes which were mantled with regolith. The importance of porewater pressure in influencing stability has already been stressed; on a slope the maximum porewater pressure occurs when the water table is coincident with the ground surface. If failure does not occur in this situation, then the slope can be considered to be stable. The slope stability model as used by Rouse and Farhan (1976) for the situation of maximum porewater pressure can be written as

$$\alpha_L = \arctan\left[\left(1 - \frac{\gamma_w}{\gamma_s}\right)\tan \phi'_r\right] \qquad (7.6)$$

where

α_L = limiting angle of valley-side slope
γ_w = density of water
γ_s = saturated bulk density of soil (from equation (7.5))
ϕ'_r = residual angle of shearing resistance

The effect of cohesion is ignored since it is approximately zero for this situation. The value of α_L was worked out using equation (7.6) for their sites and the resultant arithmetic mean was $17 \cdot 6°$.

It was then possible to compare this value with valley-side slope angles as they actually existed. At the 95% confidence level it was estimated that the population mean for valley-side slopes was

$$15 \cdot 5° \leq \mu \leq 17 \cdot 8°$$

The important result, thus, was that the valley-side slopes are at, or very near to, the mean limiting slope value, a result which has also been obtained by Carson and Petley (1970) for slopes on Exmoor and the Peak District of England. It is thought that the slopes achieved their limiting angles when periglacial processes were dominant—thus the slopes are essentially relict in form today. But Rouse and Farhan (1976) point out that such activity as constructing a road or building could re-activate movement.

7.2 Fluids

Liquids and gases possess the ability to flow and before particular properties of these states are examined it is useful to examine the nature of fluids at rest (hydrostatics) and fluids in motion (hydrodynamics).

A basic property of fluids is pressure, an understanding of which has been assumed in the preceeding chapters. Suppose the porewater pressure in a soil at a depth of 10 m had to be calculated if the water table was coincident with the surface. Visualize a column of water 10 m high and with a cross-sectional area of 1 m^2. The resultant mass on the area at a depth of 10 m would be 10 000 kg which is converted into a force when multiplied by $9 \cdot 81$ m s^{-2} (the value of g); this force in newtons ($9 \cdot 81 \times 10^4$ N) is spread over an area of 1 m^2 and thus the porewater pressure at 10 m is $9 \cdot 81 \times 10^4$ N m^{-2} or $98 \cdot 1$ kN m^{-2} if atmospheric pressure is ignored. The same type of calculation could be done to determine the atmospheric pressure at the earth's surface, though there would be the complication arising from the variability in density of the atmosphere because of differences in pressure. The method of measuring atmospheric pressure is to determine the length of a moving column supported by the atmosphere or to monitor the deformation of a partial vacuum (an aneroid barometer). It is worth noting that pressure at a particular level in a fluid at rest is constant irrespective of the shapes of the containing vessels (figure 7.16). This explains why pore pressure at a particular depth in a saturated soil is constant since variations in the pore geometry have no effect; it is the vertical difference which

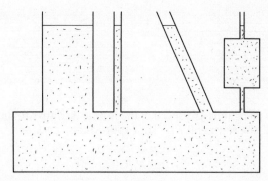

Fig. 7.16 Demonstration of hydrostatic principle.

is critical. Related to this is *Pascal's principle* which states that if the pressure at any point in an enclosed fluid at rest is changed, then the pressure changes by an equal amount at all other points in the fluid. Another classic principle which must be mentioned is concerned with the immersion of a body in a fluid— *Archimedes' principle*. A solid, when immersed in a fluid, is subjected to an upward force equal to the force resultant upon the displacement of the fluid. Suppose that a stone of granite has a mass of 10 kg and a bulk density of 2 650 kg m^{-3}. Knowing that density is obtained by dividing mass by volume, then the volume of this stone can be calculated by dividing 10 kg by 2 650 kg m^{-3} to give an answer of $3 \cdot 77 \times 10^{-3}$ m^3. If this stone occurs on a stream bed, then it will displace $3 \cdot 77 \times 10^{-3}$ m^3 of water and thus, according to the principle of Archimedes, the stone will have a mass of 6·23 kg in the stream, an apparent mass reduction of 37·7%. A stone, of course, sinks in water and this is because the stone is much denser than water. Material less dense than water will float. These principles are also directly relevant to movement within the atmosphere since a parcel of air, if it is denser than its surroundings, will sink, while if the opposite is the case, it will rise. This theme will be returned to in section 7.2.2 when gases are discussed.

In the physical environment fluids are virtually always in motion and thus principles drawn from hydrodynamics which is a specialized and advanced branch of mechanics have to be applied. It is useful to begin by examining the nature of the internal friction of a fluid—viscosity, a property which has already been mentioned in the previous section as well as in earlier chapters. Liquids and gases display viscosity, though, of course the former are far more viscous than the latter. A fluid in motion can be viewed as the combined movement of many thin layers which are individually displaced with respect to each other; the friction between these layers is the viscosity. In a river, for example, the maximum viscosity will be in the zone where there is the greatest rate of change of velocity—between the centre and the banks and bed where roughness reduces velocity.

A viscosity measure can be explained if the movement of a liquid over a horizontal surface is viewed as being analogous to the similar movement of a pack of cards (figure 7.17). The lowermost card moves the least because of the friction between it and the surface, but every other card moves with respect to the one below it to result in the second situation. Suppose F is the tangential force applied to cause the displacement, A is the surface area of the cards, l the

Fig. 7.17 Laminar flow of a fluid: analogous to the displacement of a pack of cards.

thickness of the pack and v the speed of the top card relative to the bottom one. Then when v is not too great F/A is proportional to v/l, or

$$\frac{F}{A} = \eta\frac{v}{l} \qquad (7.7)$$

where η is defined as a constant called the coefficient of viscosity. Study of equation (7.7) reveals that η is the force per unit area required to maintain unit difference of velocity between two parallel layers which are a unit distance apart. An example of dimensions for η is newton-seconds per square metre; measures in poises will also be encountered and these are equal to 0·1 of the above dimensions. Viscosities are often expressed in centipoise (10^{-2} poise) or micropoise (10^{-6} poise). In a strict sense this type of viscosity should be described as *dynamic* or *absolute* to distinguish it from *kinematic* viscosity which is obtained by dividing dynamic viscosity by density to give the dimensions area per time unit. This apparently strange measure is described by Carson (1971) as indicating the amount of interference per time unit between parallel layers of fluid in motion. Values for these coefficients are very much dependent upon temperature—for gases increases in temperature result in higher viscosity values whilst the opposite is the case with liquids (table 7.2).

Table 7.2 Dynamic viscosity values for water and air at various temperatures. (Data for water from Weast (1974, F-49) and for air, from Sears and Zemansky (1963, p. 331).)

Temperature (°C)	Dynamic viscosity of water (centipoise)	Dynamic viscosity of air (centipoise)
0	1·787	0·0171
20	1·002	0·0181
40	0·653	0·0190
60	0·467	0·0200
80	0·354	0·0209
100	0·282	0·0218

The suggestion has already been made that fluid flow may be akin to the movement of a pack of cards. This type of idealized flow is described as being *laminar* to indicate that flow proceeds by individual separate layers moving over others. Clearly true laminar flow only occurs under very special circumstances, and instead most flow includes movement within the flow distinctly different from the general direction of movement. In other words *turbulence* must occur. This may, for example, be primarily a series of movements along circular paths. The other extreme of laminar flow, then, is turbulent flow and these two types of motion in rivers are characterized by different velocity–depth profiles (figure 7.18). Prediction of flow type in

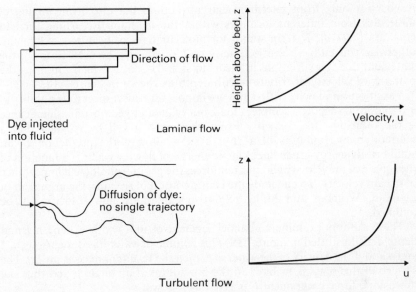

Fig. 7.18 Laminar and turbulent flow conditions. (from Carson 1971, p. 18)

streams is possible by calculation of a dimensionless parameter called the Reynolds number (*Re*)

$$Re = \frac{vd}{v} \qquad (7.8)$$

where v is velocity, d is depth and v is kinematic viscosity (Leopold, Wolman and Miller 1964). High values of *Re* indicate flow in the turbulent range whilst low values predict laminar flow.

It has been noted that a hillslope geomorphologist benefits from a basic grounding in soil mechanics; the same applies to a fluvial geomorphologist in that he must have a command of hydraulic principles. Such an approach is implicit in the classic text by Leopold, Wolman and Miller (1964). To amplify this point it is instructive to outline some aspects concerned with the velocity of streams. A resistance to flow is exerted along the water–channel interface and is a shearing stress. With laminar flow there will also be shear stress between the individual layers whilst with pure turbulent flow, shearing will take place between the components of turbulence. Shear stress (τ) within the fluid is reflected in the rate of change of velocity (v) with depth (y). Thus the equation

$$\tau = K\frac{dv}{dy} \qquad (7.9)$$

can be written where K is a coefficient. The distinction has already been made between dynamic and kinematic viscosity; it has to be noted that values for these viscosities can be obtained for two different situations. First, viscosity with laminar flow is described as being *molecular* since the interaction between the constituent layers is resultant upon intermolecular attraction. The second type, called *eddy* viscosity, arises with turbulent flow since the internal friction

E**

results not only from intermolecular attraction, but also from mixing of molecules from different localities within the medium. Returning then to equation (7.9), the K term stands for molecular viscosity if flow is laminar, otherwise it is eddy viscosity. Equation (7.9) can also be used to make the statement that it is variations in shear stress and viscosity which condition the patterns of velocity distribution with depth as shown in figure 7.18.

Variations in velocity with depth are important with respect to the ability of a stream to transport a sediment load, but of greater geomorphic significance is the value for flow velocity. From chapter 4 it will be recalled that a thermodynamic analysis of a river system allowed the prediction of an equilibrium energy grade line; kinetic energy of flowing water is a function of the square of the velocity and this reinforces the geomorphological importance of stream velocity. To conclude the consideration of streams, the argument of Leopold, Wolman and Miller (1964) in explaining flow velocity will be outlined.

These authors examine a channel segment of unit width w, length L and depth d as illustrated in figure 7.19. This volume of water has dimensions w, L and d and if the density of the water is ρ, then the resultant mass is $\rho w L d$. The surface of this mass is inclined to the horizontal at an angle β and thus the downslope force component F is given as follows:

$$F = \rho g \, w \, L d \sin \beta \qquad (7.10)$$

where g is the acceleration due to gravity. It can be readily appreciated that the angle of slope of the surface of a stream is very slight and it is thus valid to write:

$$\sin \beta \simeq \tan \beta = s$$

where s is thus the value of the tangent of the slope angle. The resisting force, equal to the impelling force since velocity is constant, is obtained by multiplying stress (τ) by the area over which the stress is applied:

$$F = \tau(2d + w)L \qquad (7.11)$$

Equations (7.10) and (7.11) can thus be combined to mean that

$$\rho g \, w \, L d \, s = \tau(2d + w)L$$

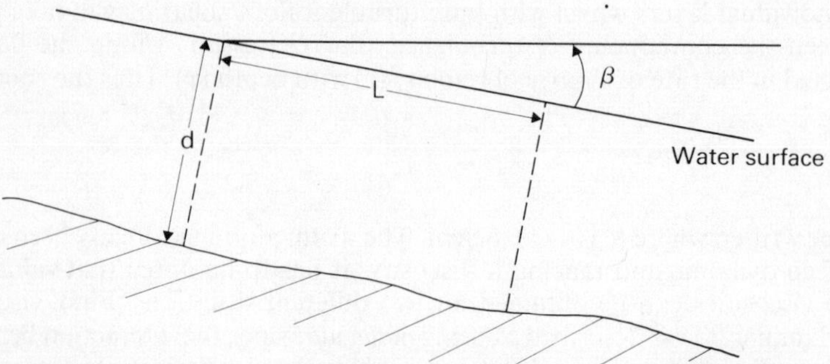

Fig. 7.19 Section along a stream showing the channel segment as discussed in the text.

Since dw is the cross-sectional area (A), then

$$\rho g\ A\ s = \tau(2d+w)$$

or

$$\tau = \rho g\ s\ \frac{A}{2d+w}$$

Now the ratio $A/(2d+w)$ is the hydraulic radius (R) so that the shear stress can be expressed as

$$\tau = \rho g\ R\ s \tag{7.12}$$

The variables on the right-hand side of this equation are hydraulic radius and slope; this latter variable has been described as the angle of slope of the channel surface, which is approximately the same as the downstream rate of loss of potential energy. Leopold, Wolman and Miller (1964) note that resistance in turbulent flow has been found by experiment to be proportional to the square of the mean flow velocity. Thus

$$\tau = k_1\ v^2$$

where τ is shear stress, k_1 a constant and v mean flow velocity. From equation (7.12) the following equation can then be written:

$$k_1 v^2 = \rho g R s$$

or

$$v^2 = k_2 \rho g R s$$

where

$$k_2 = 1/k_1$$

$\sqrt{k_2 \rho g}$ is called the Chezy coefficient, C, so that

$$v = C\sqrt{Rs} \tag{7.13}$$

which states that the mean velocity of flow is proportional to the square root of hydraulic radius and slope.

It is hoped that this brief hint at the nature of hydraulics is sufficient to indicate the importance of these principles with respect to fluvial geomorphology; clearly the interested reader must follow up these topics in a specialized text. So far concern has been with the stream as an example of a fluid, but the same type of physical approach to the analysis of sand movement by wind has been propounded by Bagnold (1954) in a classic text. However, as hinted with soil mechanics, it would be wrong to overemphasize the applicability of engineering principles to geomorphological situations, a reservation noted by Young (1972). For example, a variety of equations are available in engineering literature for predicting bedload and suspended load, but Nanson (1974) has shown that actual patterns of bedload and suspended load transportation are better explained in terms of variation in the geomorphic processes which make the sediment available to the stream. Another topic within physical geography where consideration of fluid dynamics is extremely important is coastal geomorphology. For example the nature and form of beaches are markedly influenced by the energy input—the waves. The reader is referred to Komar (1976) for an introduction to the theories of wave motions, and to wave generation, travel and breaking.

7.2.1 Liquids

Liquids, unlike crystalline solids, have no characteristic shapes, and are also almost incompressible, which distinguishes them from the other fluids—gases. The constituent molecules of liquids are continuously moving which gives liquids (and gases) the ability to diffuse; the rate of diffusion depends on such factors as the kinetic energies of individual molecules to thus give importance to the velocity of molecules as well as their mass. Velocity is conditioned to a large extent by temperature. In liquids almost continuous collisions take place between molecules whilst in gases the free path of travel of individual molecules is normally much greater than in liquids and this explains why diffusion or mixing of gases is much more rapid than for liquids.

The idea of individual molecules possessing energy and moving about within a liquid helps to explain evaporation in which certain molecules escape from the liquid to the gaseous state. In order for such movement to take place out of the liquid, the individual molecules must achieve energy levels sufficient to be able to break intermolecular bonds within the liquid. Presumably such higher energy levels are possessed by certain molecules through the random effects of collisions. Loss by evaporation clearly means a decrease of energy which will cause a drop in temperature of the liquid if it is insulated. If evaporation is aided by the movement of air over the liquid surface, then a distinct lowering of temperature can be achieved. It is also possible for a liquid to gain molecules from a gas; the dynamics of liquid loss or gain can be illustrated by considering a beaker, say containing water, enclosed in a bell jar containing oxygen (figure 7.20). Suppose that the gas in the bell jar at first contains no molecules of the type composing the liquid. Evaporation can thus proceed with water molecules migrating from the liquid into the gas. As more water molecules diffuse through the oxygen, there will be a greater chance that water molecules pass back into the liquid. Indeed the situation will evolve until the rate of evaporation equals the rate of condensation. Thus dynamic equilibrium will prevail, which means that the amount of water molecules in the beaker and the oxygen gas will be constant though there will be continual interchange between the two states. The pressure exerted by the water molecules in the

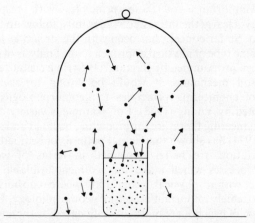

Fig. 7.20 Beaker of water enclosed in a sealed container to illustrate processes of evaporation and condensation.

oxygen is called the *equilibrium vapour pressure* which depends not only on the type of molecules, but also on the temperature.

All liquids suffer a loss if the surrounding vapour pressure is below the equilibrium value, but vaporization is most marked when boiling occurs. At this temperature the vapour pressure of the liquid is the same as the surrounding atmosphere. In this situation, vaporization is not limited to the surface of the liquid, but takes place throughout; this is possible because the vapour pressure is sufficient for bubbles of vapour to form within the liquid. As discussed in chapter 6, the boiling point of a liquid will depend on atmospheric pressure and thus it is necessary to define the *standard* or *normal* boiling point as the temperature at which the vapour pressure of the liquid is the same as one standard atmosphere (equivalent to 76 cm of mercury). The energetics of liquid/gas and solid/liquid phase changes have been described in chapter 6. Other important properties of liquids which have also been examined are surface tension (chapter 4) and viscosity earlier in this chapter.

A liquid, if pure, is composed of one substance, but such a liquid is rarely encountered in the physical environment. Instead nearly every naturally occurring liquid is a solution. In a general sense solutions are not just applicable to liquids but also to homogeneous mixtures of gases or solids. For example, the air in the atmosphere is a solution since not only is it a mixture of gases, but also it is homogeneous implying that it has distinct properties different from those of the constituent gases. But for present purposes, the discussion is limited to liquid solutions. It is essential for any quantitative analysis to be able to express in some way the concentration of gases, liquids or solids which are dissolved in a liquid. Measures of molarity, normality and molality have been described in chapters 2 and 6. It will be recalled that measurement of pH gives the concentration of H^+ ions, another example of solution strength. The solution of substances in a liquid has an effect on the properties of the liquid; in particular the magnitude of the change in liquid properties is a direct function of solute concentration. For example, salt water freezes at a lower temperature than pure water. Similarly there will be a slight elevation of the normal boiling point temperature with an aqueous solution in contrast to pure water (figure 7.21). Such differences can be explained when it is remembered that the effect of solution is to decrease the free energy (chapter 4). The explanation of osmosis is also intimately involved with these principles (chapter 4). Many of the characteristics of liquids have been described in earlier chapters; in order to summarize some of the attributes, it is useful to describe some of the key properties of that ubiquitous environmental liquid—water.

Water has its normal freezing point and normal boiling point at 0 °C and 100 °C respectively. The open structure of ice means that ice is less dense than water in the liquid state; another unusual property is that water has its maximum density ($1 \cdot 0000$ g cm^{-3}) at $3 \cdot 98$ °C. Of major environmental significance is the comparatively high heat capacity of water which means that it absorbs or emits much heat to change its temperature and thus large water masses have a very significant moderating effect on local climates. The solvent power of water is well recognized, but it is also important to stress that many substances are virtually insoluble in it. In general, for example, many hydrocarbons can be considered to be insoluble in water since the strength of hydrogen bonding in the water molecules is too strong to permit interaction

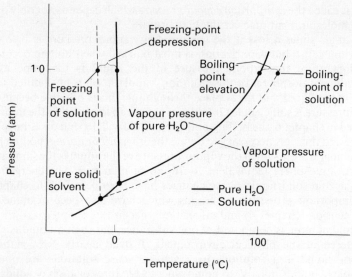

Fig. 7.21 Phase diagrams for an aqueous solution of a non-volatile solute (dashed line) compared to that of pure water (solid line). (from Anderson, Ford and Kennedy 1973, p. 322)

between the water and hydrocarbon molecules. In contrast ionic substances dissolve in water, albeit in very different amounts; solution is favoured because of the interaction between dipolar water molecules and ions. The energy required for such dissociation will depend upon the nature of the ionic lattice. Following dissociation, water molecules are attracted to individual ions or to other charged bodies such as clay surfaces. The result is a decrease in free energy of the water molecules. This is reflected, for example, in a phenomenon called the *heat of wetting* which arises when clays adsorb water molecules; the result is a liberation of energy consequent upon the decrease of free energy. A similar process occurs when ions and water molecules are bonded (hydration).

A final point worthy of mention is that naturally occurring water is made up of a mixture of hydrogen and oxygen isotopes. In particular, average values for the percentage composition of hydrogen in water are given as follows:

$$_1^1H \quad 99 \cdot 985\%$$
$$_1^2H \quad 0 \cdot 015\%$$

whilst for oxygen the corresponding values are:

$$_8^{16}O \quad 99 \cdot 759\%$$
$$_8^{17}O \quad 0 \cdot 037\%$$
$$_8^{18}O \quad 0 \cdot 204\%$$

(Anderson, Ford and Kennedy 1973 p. 334). The $_1^1H$ isotope is called protium whilst $_1^2H$ is called deuterium; a third isotope of hydrogen also occurs ($_1^3H$)—tritium. The fact that water is a varying isotopic mixture has been utilized for identifying variations in temperature through time since the ratio of the two oxygen isotopes is temperature dependent. This has led to the analysis of deep ice cores taken from such areas as Greenland and Antarctica; ice at lower levels is clearly of greater age than surface ice and this has encouraged research workers such as Dansgaard *et al.* (1969) to determine the $^{18}_{8}O/^{16}_{8}O$ ratio for many depths in order to postulate variations in temperature through time.

7.2.2 Gases

A gas, unlike a solid or liquid, is almost all empty space with individual molecules only taking up a very small amount of space. These molecules may all be identical or they may be different isotopes of one or more elements. Individual molecules are constantly moving, colliding with each other as well as with the walls of the container: these collisions on the container wall result in the gas exerting a pressure whilst the rate of movement of gas molecules is a function of temperature. Experimental proof for the constant movement of molecules was first presented by Robert Brown who observed under a microscope the irregular movements of tiny particles when suspended in a liquid or gas: such a phenomenon was subsequently called Brownian motion. This provided the basis to the *kinetic theory of matter* which has as one basic postulate that molecules are always in motion. This theory can easily be applied to explain the properties of gases, but like all theories, certain assumptions must be made. Thus an ideal gas is defined to accord with the kinetic theory and it is assumed that the space occupied by molecules is so small that it can be ignored: molecules are treated as points which move about with a rapid, random, straight-line motion and no force exists between these points; it is also assumed that all collisions are perfectly elastic and that there is no net loss of energy resultant upon collisions. In practice no real gases behave as though they are ideal gases, but for certain conditions, the similarity is marked. This is the justification for describing the gas laws which will allow a general relationship between volume, pressure, temperature and number of moles of a gas sample to be obtained.

Boyle's law

Robert Boyle in 1660 described the relationship which he obtained between pressure and volume. In particular, he found that if the temperature of a gas was kept constant while its volume varied, then the pressure varied in such a way that the product of pressure and volume remained the same. Thus if P is pressure and V is volume,

$$PV = k_1 \qquad (7.14)$$

where k_1 is a constant, if temperature and mass are unchanging. An ideal gas perfectly accords with this law. A consequence of Boyle's law is that if these special conditions apply, a doubling of volume will result in a halving of pressure.

Charles' law (or Gay-Lussac's law)

This law describes the relationship between volume and temperature when pressure is kept constant. In detail, the volume occupied by a given mass of gas at different temperatures within moderate ranges is directly proportional to the absolute temperature if pressure is kept constant. Thus

$$V = k_2 T \qquad (7.15)$$

for constant pressure and mass where V is volume, T absolute temperature and k_2 a constant. This relationship is illustrated in figure 7.22; the graph is extended to the temperature axis to indicate that the gas, if it were to remain a gas, would occupy no volume at absolute zero (0 K).

Boyle's and Charles' laws can be linked to express a combined gas law. In this situation pressure (P), volume (V) and absolute temperature (T) vary, but the mass of the gas remains constant. The formal statement of the combined gas law is as follows:

$$\frac{PV}{T} = k_3 \qquad (7.16)$$

at constant mass and where k_3 is a constant.

The only limitation in the use of equation (7.16) is that the mass of gas must remain constant. Clearly an equation which included a variable quantity of gas would be more versatile in application. In order to do this, a brief return needs to be made to Avogadro's number. It will be recalled that one gram atomic weight of any substance contains $6 \cdot 02 \times 10^{23}$ molecules—the quantity of one mole. One mole of gas molecules, irrespective of the type of gas, at $273 \cdot 15$ K and at one atmosphere pressure occupies $22 \cdot 414$ litres. Thus the volume (V) occupied by a gas at constant temperature (T) and pressure (P) is directly proportional to the amount of gas, i.e.

$$V \propto n \text{ at constant } T \text{ and } P$$

where n is the number of moles. Similarly Boyle's law can be written as

$$V \propto \frac{1}{P} \text{ at constant } T \text{ and } n$$

and Charles' law as

$$V \propto T \text{ at constant } P \text{ and } n$$

These proportionalities can be combined as follows:

$$V \propto \frac{1}{P} T n$$

or

$$V = R \frac{1}{P} T n \qquad (7.17)$$

where R is a constant and is known as the universal gas constant. Equation (7.17) is written in the more common form

$$PV = nRT \qquad (7.18)$$

and this equation is the *equation of state* or the *perfect gas law* for an ideal gas. Since any ideal gas occupies $22 \cdot 414$ litres at standard conditions, the value of R can be calculated as follows:

$$R = \frac{PV}{nT} = \frac{(1 \cdot 000 \text{ atm}) (22 \cdot 414 \text{ l})}{(1 \cdot 000 \text{ mol}) (273 \cdot 15 \text{ K})}$$

$$= 8 \cdot 2057 \times 10^{-2} \text{ l atm K}^{-1} \text{ mol}^{-1}$$

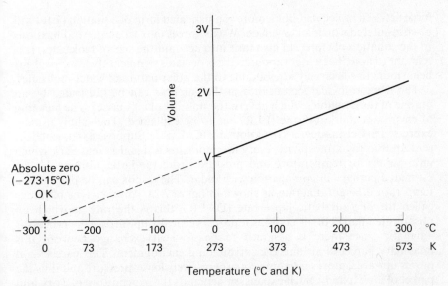

Fig. 7.22 Relationship between volume and temperature with constant pressure. (after Anderson, Ford and Kennedy 1973, p. 29)

An alternative value for R, the universal gas constant, is $8\cdot3144\,\text{J K}^{-1}\,\text{mol}^{-1}$. Knowledge of equation (7.18) as well as the gas constant allows the calculation for any gas of changes in one property with respect to other properties. An example should help to clarify these principles.

Suppose $0\cdot5910$ g of a pure gas occupies $0\cdot9$ l at a temperature of 298 K and at a pressure of $1\cdot00$ atm. Equation (7.18) can first be used to calculate the number of moles present:

$$n = \frac{PV}{RT} = \frac{(1\cdot00\ \text{atm}) \times (0\cdot9\ \text{l})}{(0\cdot082057\ 1\ \text{atm mol}^{-1}\ \text{K}^{-1}) \times 298\ \text{K}}$$

$$= 0\cdot0368\ mol$$

Care should always be taken to ensure that the units are consistent so that in this case, cancellation of units leaves n with the dimension of moles. Next the molecular mass can be calculated by dividing the mass of the gas ($0\cdot5910$ g) by the number of moles ($0\cdot0368$ mol) to give the answer $16\cdot06\ \text{g mol}^{-1}$ to indicate the gas must be oxygen (example adapted from Anderson, Ford and Kennedy 1973, p. 270).

As already mentioned actual gases only approximate to these ideal conditions. In physical geography, in particular in meteorology, an understanding of the equation of state is fundamental to many phenomena rather than an ability to make modifications to correct for deviations from the ideal conditions. Nevertheless, it is interesting to note that the method assumes no attraction between molecules in a gas, whilst already described in this chapter has been the nature of van der Waals' forces which would be present between molecules. At specific instants molecules can be attracted to each other because of adjacent opposite charges. At high temperatures, the rate of movement of individual molecules is such as to cause a decrease in this net force between molecules, but a more sophisticated form of equation (7.18) will be encountered which takes van der Waals' forces into account. Modifications to the equation of state can also take into account the size of molecules. It is relevant to note that water vapour as a gas deviates significantly from ideal gas behaviour; this is largely attributable to the polar nature of water molecules.

The meteorological significance of the gas laws can be illustrated by an outline of the dynamics which govern the stability of air masses. The meaning of environmental lapse rate (ELR) has to be explained. The rate is usually expressed in Celsius degrees per kilometre (C° km^{-1}). Suppose a radiosonde is sent up into the atmosphere, every few moments a signal is sent back which gives values for temperature and pressure: from the latter altitude can be estimated perhaps in conjunction with radar. The results can be plotted in a temperature/height diagram as shown in figure 7.23. Another type of rate is called the dry adiabatic lapse rate (DALR); this is the rate at which an unsaturated parcel of air cools as it moves vertically upwards—the process is adiabatic since there is assumed to be no heat exchange between this imaginary parcel of air and the surrounding environment. This parcel, as it moves upwards, moves into areas of progressively lower pressure and thus the parcel of air expands. Such expansion demands the expenditure of work and this is reflected in a drop in temperature. In effect the parcel is a closed thermodynamic system and the adjustments of temperature, pressure and volume are explainable in terms of the equation of state for an ideal gas. If the vapour pressure of the parcel of air is below the saturated value, then the parcel

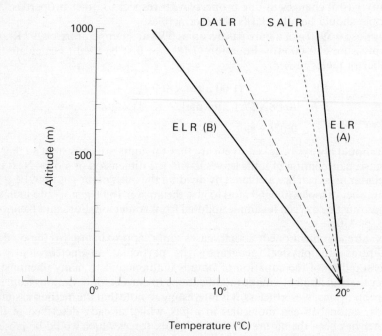

Fig. 7.23 Temperature–height diagram to illustrate a stable air mass (line A) and an unstable air mass (line B). The dry adiabatic lapse rate (DALR) and saturated adiabatic lapse rate (SALR) lines are also shown.

will cool at a constant rate which is approximately 10 C° km^{-1}. Lines are drawn on figure 7.23 to show the nature of the dry adiabatic lapse rate.

Air is defined as stable if, on displacement, it returns to its original point of disturbance; similarly it is described as unstable if, on displacement, it accelerates away from the point. These principles are of fundamental importance with respect to the nature of air masses. If unsaturated air is being considered, the stability or otherwise of an air mass can be determined by the following argument. The temperature of a parcel of air at any altitude is determined assuming it cools adiabatically and is compared with the actual environmental temperature at that point. For example, in figure 7.23, consider the ELR labelled A: if an unsaturated parcel of air is imagined to rise from the ground where the air temperature is 20 °C, at 1 km the temperature of the parcel is 10 °C whilst the environmental temperature is 18 °C so that the parcel will be denser than the surroundings and will thus sink back to the surface—a stable situation. If, on the other hand the ELR had the form as indicated by B in figure 7.23, then the parcel would be warmer than the surroundings and would continue to rise—a situation of instability.

The situation is a little more complicated should water vapour become saturated. By this it is meant that the vapour pressure has reached its equilibrium value for a particular temperature. Humidity is the term which is often used to describe the quantity of water vapour present (chapter 6). In particular, a measure of relative humidity is most useful since this dimensionless value expresses as a percentage the amount of water vapour actually present at a particular temperature compared to the amount which would be present if saturated conditions existed. Thus if water vapour in air had achieved equilibrium pressure, then the air mass would have a relative humidity of 100%.

The previous argument about an imaginary rising air parcel can be repeated, but this time the air is saturated. As the parcel rises, lower pressures lead to expansion of the parcel which causes the expenditure of energy resulting in a drop in temperature. Since the water vapour is already at equilibrium for a particular temperature before descent, then the only possible response at lower temperatures is for the parcel to reduce its water vapour pressure by condensation taking place. This process yields latent heat of condensation which has the effect of reducing the rate of temperature decline with altitude. The consequence is that the saturated adiabatic lapse rate (SALR) is significantly less than the DALR. In marked contrast to the DALR, the SALR is not a constant, but varies from about 4 C° km^{-1} at high temperatures to a value tending towards the DALR at very low temperatures. Such variability can be explained by noting that at very low temperatures, water vapour pressures are very slight and any further lowering of temperature will yield very little heat from condensation.

These pressure–volume–temperature relationships are thus integral to an analysis of an air mass to determine its stability. In figure 7.23 a SALR line has been drawn assuming it to be constant. Using the argument as advanced before, the air mass represented by the ELR line (A) is still stable if the air mass is saturated and similarly the other air mass (B) is unstable. These two situations are described as absolute stability and absolute instability respectively. If an ELR curve had occurred between the DALR line and the ELR line, then the air mass would be conditionally unstable. The

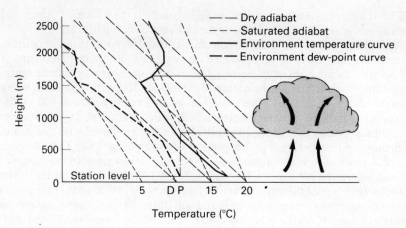

Fig. 7.24 Fair weather cumulus on a temperature–height diagram. (from Miller and Parry 1975, p. 78)

meteorological significance of these principles must be followed up in a suitable text such as Barry and Chorley (1976) or Miller and Parry (1975); suffice it for present purposes to discuss the meteorological situation as presented in figure 7.24. The surface temperature is 17 °C and the variations with height are described by the ELR line. From the surface a DALR line is drawn and this intersects a vertical line drawn from the surface dew-point, the temperature to which air must be cooled at a particular pressure and moisture content so that the equilibrium vapour pressure of water occurs. At around 700 m saturated conditions prevail to give a distinct cloud base. Above this height an imaginary parcel of air would cool at the SALR and eventually this line intersects the ELR at an altitude of about 1650 m. Below this height the parcel of air would always be warmer than the surrounding environment meaning that unstable conditions prevail. Above this height the reverse is the case leading to a stable situation. Thus the upper and lower levels of a fair weather cumulus cloud can be explained in terms of convection processes within a distinct altitudinal range.

The aim of this chapter was to demonstrate the nature and properties of solids, liquids and gases. In no way was the intention to try to offer a comprehensive description of individual properties relevant to physical geography; instead the aim was to show by example how the fundamentals of chemistry and physics assist with a comprehension of these properties. A firm command of basic science allows topics to be approached from first principles. Thus the reader who is confronted with a climatological text which analyses atmospheric motion ought not to be dismayed if he understands the nature of the forces involved and how they can be resolved as well as the other principles associated with fluid motion. Another general concluding point worthy of stress is that in the physical environment, the three states of matter, solid, liquid and gas, rarely occur as separate entities. Nearly every segment of the physical environment consists of some combination of these states. Thus not only is there a need to understand their individual properties, but also their characteristics in combination as well as the mechanisms which lead to equilibrium situations.

8
Errors and the reporting of results

The preceding chapters have outlined the principles of science relevant to physical geography. Without such a basic knowledge, investigation of the physical environment can only be at a fairly elementary and descriptive level. Physical geographers, or environmental scientists must be able to analyse problems using appropriate concepts drawn from physics or chemistry; mathematical ability is also an essential skill. The physical geographer, like any scientist, soon becomes involved with experimentation and again, physics and chemistry are often integral not only to the identification of the problem, but also to the technical execution of the experiment.

Experimentation is essential for the advancement of a field science like physical geography. Thus, in a book dealing with science for physical geographers, an appropriate way to conclude is to focus attention on the outcome of experimentation—the results. Experimentation necessitates measurement and it is impossible for such an operation not to include errors. The assessment of errors is essential before any results can be compared. Emphasis is given in this chapter to the nature and calculation of errors, but before this is tackled, the distinction between accuracy and precision must be made clear.

The object of any measurement is to make an estimate of the *true value* which in a strict sense is unobtainable. A true value would only be obtained if the limit of an infinite series of observations were made, all carried out under the same conditions, but this is a practical impossibility. However, just because an objective is unobtainable does not mean that it does not exist; in fact one of the assumptions of measurement is that a true value does exist. Griffiths (1967) suggests that a true value can be considered as the bull's eye in a circular target (figure 8.1). The object of measurement is to make a good estimate of its value. The results of rifleman A are relatively accurate since they are regularly scattered round the bull's eye but are imprecise because of their degree of scatter. In contrast rifleman B shoots with far greater precision, but is inaccurate. Accuracy is deviation from the average of a scatter of shots to the centre (true value) whilst precision is the degree of scatter between the shots of one rifleman. As can be seen from the attempts of rifleman C, he is both inaccurate and imprecise. It is clear that accuracy and precision need not be dependent—the improvement in precision of measurement need not necessarily be paralleled by a corresponding increase in accuracy.

It is important to appreciate that the arithmetic average of a set of measurements is not necessarily the same as the true value. In practice it is often difficult to determine if there is a difference between a 'statistical true value' and an 'actual true value'. Such a difference, if it occurs, results from the

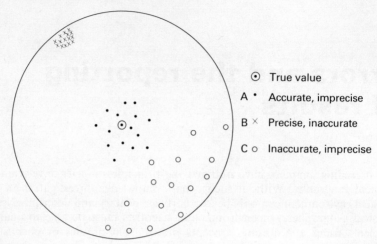

Fig. 8.1 Rifle target with different shot patterns to indicate the difference between accuracy and precision. (after Griffiths 1967, p. 4)

introduction of *bias* to the measurements. The results from rifleman B in figure 8.1 are very much biased in one direction. This could arise from a faulty setting on the gun or consistent error on the part of the marksman. The difference between bias and precision is shown in figure 8.2. Measurement technique A has low precision but the arithmetic average of the results is the same as the true value. In contrast, measurement technique B is far more precise, but a distinct bias is evident. The precision of both results can be improved by increasing sample size, but bias need not be influenced. Factors influencing precision and bias will be discussed in the following section, but before any attempt is made to reduce bias it is important to first ask if this is necessary. If

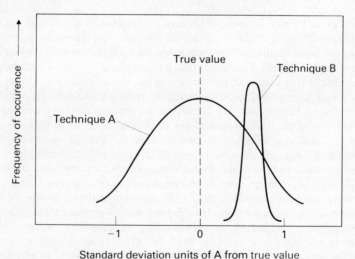

Fig. 8.2 Comparison in distribution of results from a technique of low precision (A) with a precise one (B) which is biased.

measurements are to be made for comparison between themselves, then some bias may be acceptable. For example, say measurements of soil moisture content are made along a slope profile, then interest is in the *comparative* values. An approximately equal bias in the same direction for each moisture content determination would not necessarily influence any trend. If the results were to be compared to those of a different survey, then absolute determinations would be required with controlled bias.

8.1 Types of errors

Bias and lack of precision and accuracy are a reflection of a variety of types of error. Theoretically any result should give some indication of total error—the sum of all the errors involved in the procedures. In practice it is often difficult to estimate total error, but nevertheless it is important in any analysis to be aware of all types of error and to have some idea of their magnitudes. Errors are introduced to measurements either separately or in combination. Ackoff (1962) summarizes errors resulting from: (1) the observer; (2) the instrument used; (3) the environment; and (4) the thing observed. Sampling error also has to be considered.

(1) Observer Error

The human element in measurement necessarily introduces errors through faulty reading of instruments, failure to follow a technique exactly, lack of concentration or by making a slip in recording a result. The precision and bias of measurements will vary according to the operator and such differences are a reflection of *operator variance*. For example, imagine that two students were instructed to weigh out a large number of samples of sand each weighing 10·000 g and then the results were checked on a perfectly calibrated automatic digital balance. The results are plotted on a frequency graph (figure 8.3) and as can be seen, student A has produced values with a marked peak over 10·000 g and with little spread on either side, whilst with student B, not only is a large element of bias evident in his results, but also a wide scatter. Bias and spread can also vary for one observer with time. Can such errors be assessed in practice? With measurements which can be made time and time again, a distribution curve can be drawn as in figure 8.3. But bias cannot be identified by this procedure—student B could continue weighing out samples but his method would still result in a biased result unless the error in his technique was identified. Of course the actual bias of a result can be precisely determined if the true value is known. Thus in order to correct for observer bias either several observers are used, or the results from one observer are checked with reference to some standard. Chorley (1958) has carried out experiments with students to examine variation in morphometric measurements of drainage basins. With measurements of areas, stream lengths and gradient he found no differences, but operators varied in recognition of first-order channels.

A more difficult situation arises when it is possible to make only one measurement of an attribute because the technique involves alteration of the attribute or the measurement is fixed to time. Consider the determination of loss-on-ignition; an experiment cannot be reapeated on the same sample since the technique involves destruction of the organic matter. If bias must be strictly

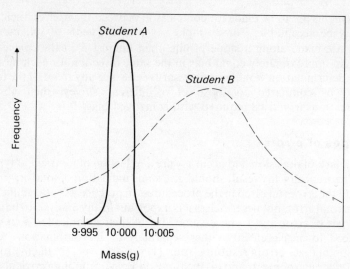

Fig. 8.3 Two distributions to illustrate marked differences in observer errors between two students.

controlled, groups of random samples are given to different operators and by estimating the contribution which sampling error makes to the total error operator error can be determined (Ackoff 1962).

(2) Instrument error

This type of error is self-explanatory; instruments, like human operators, can give biased or inconsistent results. The techniques for isolating such instrument errors are the same as with operator errors. The tendency with experimental work in physical geography is for increasing use of more sophisticated instrumentation. The advantages of automated testing techniques are obvious in minimizing human effort and error, but there is the corresponding increase in the likelihood that instruments may fail or that spurious results may not be recognized.

(3) Environmental error

The condition under which measurement takes place may well influence the results. Often allowance can be made for such effects. For example the silt and clay content of a sample can be measured by monitoring the change of density of a suspension with time; corrections have to be introduced if there are temperature changes since these will influence the density of water, the medium in which the silts and clays are sedimenting. Sometimes the rate of change of an environmental factor may be used for predicting some other attribute. For example, a surveying aneroid barometer is used for quick determination of altitude based upon differences in atmospheric pressure. Clearly such determinations are only possible when meteorological conditions are settled, otherwise complex corrections are necessary (Sparks 1953).

(4) Error due to the nature of the observed phenomenon

This arises when the measurement technique influences the results, and a good example is the measurement of soil creep. Flexible tubes may be inserted into the ground and the deformation of these tubes monitored. But the installation of the tubes and their presence may well influence the resultant values. Similar problems arise with questionnaire surveys when questions can easily give biased answers. As with all experimental work the best way of minimizing this type of error is the application of common sense and practical ingenuity.

(5) Sampling error

This type of error is rather different in character from the preceding ones since its magnitude, at least in part, results from the other errors. This can be explained by considering the nature of sampling which has to be employed so often in physical geography. A distinction needs to be drawn between measurements being made of exactly the same unchanging feature and measurements of a feature which cannot be kept identical. As an example of the former, suppose the height of a raised beach above Ordnance datum has to be determined. The topographical survey can be carried out several times. The assumption is that there is one particular true value which is estimated by several measurements. It is very unlikely that the results will be identical and the degree of scatter will reflect the occurrence of several types of errors including random effects. In practice, such a topographical survey would only be carried out once since the method would incorporate error checking procedures. It is important to appreciate that statistical analysis of repeated measurements cannot predict the total error or uncertainty of the mean or corrected mean value. Campion *et al.* (1973) draw the distinction between the uncertainty of a measurement resulting from random uncertainty in contrast to systematic uncertainty; the former is assessed by statistical procedures whilst the latter is evaluated by experimental techniques. For present purposes the limitations of repeated measurement for estimating uncertainty should be noted.

In physical geography it is often impossible to make more than one measurement of the same phenomenon. It is impossible to obtain two or more identical soil or water samples for analysis; a velocity measurement in a stream can never be exactly repeated since the same water cannot be organized to pass the blades of the current meter again. So the variability in results from these types of measurements is caused not only by the errors as discussed above, but also by differences in the true value of the property being determined. In general these differences are consequential upon sampling in space or time. Thus an estimate of sampling error for this type of data will indicate component errors as well as variability in the phenomenon being measured. An example seems appropriate to further illustrate these important principles.

Imagine that all the sediment on the bed of a particular part of a stream is collected and taken back to the laboratory where the clay content is to be measured. Suppose the population consists of one large bag of sediment containing several kilograms. The clay content will be a percentage value of the mass of clay in the bag to the total dry mass of the material. One true value exists and the aim of laboratory analysis is to make a best estimate of that percentage. The pipette method (Bascomb 1974) is a standard method for determining clay content; the technique itself is of no importance to the present argument, but what is very relevant is that only 10 g of the sediment is required for the analysis. In other words, 10 g must be extracted from the large bag of material. If small quantities were poured out of the large bag the results would be very biased since the larger fragments would emerge first; instead a technique must be used to result in small representative samples. The common method is to use a sample divider or riffle box; the bag of material is emptied into a hopper at the top of the divider and alternate chutes divide the material into two boxes. The contents of one box can then be subdivided in the same manner and the process continued until the necessary size of sample is obtained. Suppose that fifteen samples of 10 g each have been obtained of this river sediment. Imagine that is is possible to analyse these samples for clay content without any operator or instrument error, then the differences between the fifteen results would be a sole reflection of differences in sample values— *sampling error* in a strict sense. If the riffling process was efficient in producing similar representative samples, then the variation in the results should be small. If the sampling procedure is biased, say by favouring the addition of coarser material, then the mean of the sample results will be different from the true value for the whole bag.

8.2 Statistics and the representation of random uncertainty

The consequence of all these various types of errors is that they are cumulative and thus account for the difference between the true and measured value. In practice, of course, the true value cannot be determined and one method is to assess variability in results by statistical techniques. Thus it is necessary to summarize some basic concepts in statistical analysis; readers without any statistical training are urged to refer to texts which introduce statistical methods (Gregory 1973, Hammond and McCullagh 1974).

With one measurement of an attribute, no statement can be made about uncertainty limits on statistical grounds. As the number of observations of the same attribute increases, the magnitude of uncertainty can be considered on the basis of the variability in results. The effect of increasing sample size and decreasing class width is illustrated in figure 8.4. For the imaginary situation of an infinite sample size and class width almost zero, a smooth curve is obtained. If no biased errors are included then this curve will be symmetrical about the arithmetic mean ($\bar{x}$) and the shape of the curve will be a reflection of measurement variability. The mean is calculated simply by adding up all the measurements and dividing by the number (n). If there are five measurements then

$$\bar{x} = \frac{x_1 + x_2 + x_3 + x_4 + x_5}{5}$$

where x_1 to x_5 are the individual measurements. A neater way of expressing this calculation is to write

$$\bar{x} = \frac{\sum_{i=1}^{5} x_i}{5}$$

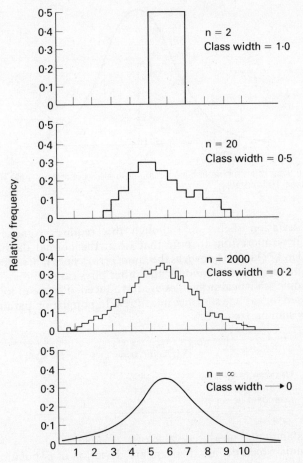

Fig. 8.4 Effect of increasing sample size and decreasing class width. (from Bragg 1974, p. 47)

The general formula for the mean then is

$$\bar{x} = \frac{\sum_{i=1}^{n} x_i}{n} \tag{8.1}$$

where $\bar{x}$ is the mean, n the number of observations, and x_1 to x_n the individual values. The mean ($\bar{x}$) in this example is an average of *sample* values; when the average of all values in a population is computed the mean is represented by μ.

In order to describe the variability of values, information is required on the nature of the distribution curves as shown in figure 8.3. The estimation of variability would be extremely difficult if a different type of curve had to be used for each problem. Instead it has been found that a very limited number of curves are applicable to a large number of situations. The best known and most widely used type is the *normal* or *Gaussian distribution*. For present purposes the relevant point to stress is that any normal curve is fixed by the mean and the

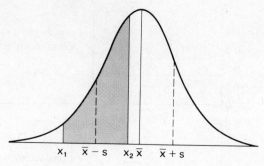

Fig. 8.5 The area under a normal curve as a measure of probability. (after Hammond and McCullagh 1974, p. 99)

standard deviation (figure 8.5) though this requires some explanation. However, it is interesting to note that when the normal distribution was introduced in 1773 it was known as the *law of errors* since it described very well observational errors in astronomy and other physical sciences.

Before some statistical principles are introduced, it is useful to distinguish notation used for sample statistics in contrast to population parameters. This notation is summarized below:

	Population parameter	Sample statistic
Number of variates	N	n
Mean	μ	$\bar{x}$
Standard deviation	σ	s

A variate is an actual measured value of a variable.

The formula which defines the standard deviation of population data is

$$\sigma = \sqrt{\left(\frac{\sum_{i=1}^{N}(x_i - \mu)^2}{N}\right)}$$

(8.2)

The corresponding formula for sample data is

$$s = \sqrt{\left(\frac{\sum_{i=1}^{n}(x_i - \bar{x})^2}{n}\right)}$$

(8.3)

As will be discussed, the term n in the denominator of equation (8.3) is often replaced by $(n-1)$; this is necessary when dealing with samples of less than 30.

Computation of s using equation (8.3) is straightforward if a calculator with an additive memory is used. An alternative formula is

$$s = \sqrt{\left[\frac{\sum_{i=1}^{n}x_i^2}{n} - \left(\frac{\sum_{i=1}^{n}x_i}{n}\right)^2\right]}$$

(8.4)

The method of calculation using this formula can be demonstrated by an example. Consider the large bag of sediment which was discussed with reference to sampling error. Suppose that the clay contents of fifteen samples of this material have been determined as follows:

Variates (% clay), x_i	Variates squared x_i^2
21·3	453·69
21·5	462·25
21·2	449·44
21·4	457·96
21·5	462·25
21·3	453·69
21·2	449·44
21·4	457·96
21·3	453·69
21·6	466·56
21·7	470·89
21·1	445·21
21·4	457·96
21·4	457·96
21·6	466·56

$$\sum_1^{15} x_i = 320\cdot9 \quad \sum_1^{15} x_i^2 = 6865\cdot51$$

$$n = 15 \quad \bar{x} = 21\cdot39$$

Substitution into equation (8.4) gives

$$s = \sqrt{\left[\frac{6865\cdot51}{15} - \left(\frac{320\cdot9}{15}\right)^2\right]}$$

$$= \sqrt{(457\cdot701 - 457\cdot675)} = 0\cdot161$$

(neglecting the correction for a small sample size)

The standard deviation has the same dimension as the data items; in the worked example above, s happens to be dimensionless since the clay content is expressed as a percentage. Calculation of the standard deviation thus gives a measure which expresses the variation in results—a reflection of their precision.

An integral part of experimental work is a statement of the range within which the true value is most likely to occur. For example the average of the fifteen clay values is 21·4%; how confidently can the statement be made that the true value lies between say 21·35% and 21·45%? Intuitively it can be proposed that greater confidence can be expressed with a wider range. In effect a statement of the following form is required: clay content of bed of stream is 21·4±0·3% with a probability of 95% of being correct. In other words a 5% or 1 in 20 chance of being wrong is accepted. In order to work out errors of the mean with their associated probabilities, consideration must return to the normal curve.

As already stated it has been found that when a large number of determinations are made of the same measurement, the results tend to be distributed in a normal manner. In other words, the characteristic symmetrical bell-shaped curve centres over the mean and the curve tails off in both directions in an asymptotic manner. If samples are drawn from a normally distributed population, then the sampling of the distribution will also be normal. The application of the normal distribution to experimental problems is best illustrated by an example, but before this is possible some properties of the normal curve must be summarized. A lucid and more detailed account is given by Hammond and McCullagh (1974). The normal curve is symmetrical, bell-shaped and is a probability distribution. This latter point requires some explanation. Consider a normal distribution as in figure 8.5; $\bar{x}$ is the mean of the samples and s is the standard deviation. If more samples were to be drawn from the same population, then the probability of these values falling in particular ranges can be predicted from the normal curve. Suppose it was required to know the probability of another sample having a value between x_1 and x_2 as indicated in figure 8.5. Since the normal curve is a probability distribution, the total area under the curve must equal the probability of a value falling within the complete range—in other words 1, which means absolute certainty. Thus the probability of this additional sample having a value between x_1 and x_2 corresponds to the area under the curve for that range, the shaded area in figure 8.5. Probability is expressed on a scale running from 0 (absolutely no chance) to 1 (absolute certainty) and such a range can also be stated in percentage terms. Important areas under the normal curve are those within one, two and three standard deviations from the mean. For example, the probability of a value falling within one standard deviation of the mean ($\bar{x} \pm s$) is 0·683; corresponding values for $\bar{x} + 2s$, and $\bar{x} \pm 3s$ are 0·9545 and 0·9973 respectively.

The analysis of penetrometer measurements illustrates not only the required statistical procedures, but also some of the practical difficulties of dealing with field data. A penetrometer is an instrument which is used to measure the resistance to penetration of a cone which is screwed on to the end of a steel rod and is pushed down into the ground whilst recording the downward force required by a gauge. One method of measurement is to push the cone into the ground at a uniform rate and to record the force required to achieve fixed depths. A simpler method involves the application of a fixed force and measurement of the depth of penetration of the cone; it is this latter method which was used to determine the resistance to penetration of beach sand at Llangrannog, Wales. A square metre grid was laid out on damp sand about 5 m away from the sea and a hundred penetrometer depth measurements were made using a pressure of 200 kN m^{-2}. On a number of occasions however the limitation to penetration was not only dependent on sand resistance, but also on the presence of pebbles in the sand. The results are presented in table 8.1. The striking feature of these results is the long tail in the distribution caused by the shallow depth determinations due to pebbles.

In order to assess the error say of the average depth of penetration, a check must be made to establish if the data have an approximately normal distribution. Various statistical techniques can be used, but the most useful method is to plot the data on probability graph paper with a probability scale on the Y-axis and usually an arithmetic scale on the X-axis (figure 8.6). In order to plot data on such paper, occurrences must be accumulated and expressed

Table 8.1 Frequency of occurrence of depths obtained in sand by a penetrometer with a pressure of 200 kN m^{-2}.

Depth range (cm)	Frequency	Cumulative frequency
26·00–26·49	2	2
26·50–26·99	0	2
27·00–27·49	1	3
27·50–27·99	0	3
28·00–28·49	0	3
28·50–28·99	2	5
29·00–29·49	2	7
29·50–29·99	1	8
30·00–30·49	2	10
30·50–30·99	2	12
31·00–31·49	1	13
31·50–31·99	4	17
32·00–32·49	8	25
32·50–32·99	10	35
33·00–33·49	21	56
33·50–33·99	26	82
34·00–34·49	12	94
34·50–34·99	5	99
35·00–35·49	1	100
Total	100	

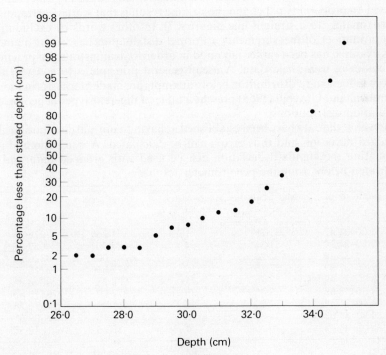

Fig. 8.6 Penetrometer data plotted on probability paper.

on a percentage basis. The cumulative frequency data for the penetrometer results are presented in table 8.1. These are also percentages since the total number of measurements was one hundred. The first point to plot has an X-value of 2 and a Y-value of 26·49 (the upper limit of the first depth range). The second point has corresponding values of 2 and 26·99, the third point values of 3 and 27·49. All the data for the depth classes are thus plotted and the data can be taken to be normal if the result is approximately linear. With the beach measurements, this is clearly not the case. Instead the scatter seems to have two trends in gradient and these seem to merge between 31 and 32 cm. With such a result the usual reaction is to attempt to normalize the data by choosing an appropriate transformation of the values. Commonly, logarithms of values are used as this reduces the relative magnitudes of large values. Alternatively values in the upper parts of a scale can be expanded. For this, powers of numbers would be taken. The only way to proceed is by trial and error. However, it is often useful to consider factors which influence a non-normal distribution before many hours are spent trying different transformations. In the case of the beach data, the skewed nature of the distribution is probably due to the presence of stones in the beach profile. If the aim is to determine the penetrability of sand then one solution is to disregard measurements when stones were encountered. In practice this cannot always be done; an alternative strategy based on the nature of the scatter in figure 8.6 is to exclude all measurements less than 31 cm assuming these were all caused by stones. It is possible that a few measurements greater than 31 cm have been limited by stones, but these are probably few in number. When results of less than 31 cm are excluded, with new percentages computed, a new scatter of points can be plotted on probability paper and the striking result is that scatter corresponds approximately to a straight line (figure 8.7). In other words by excluding a certain number of measurements a normal distribution has been obtained. This example has been rather laboured in order to demonstrate the practical difficulties of using field data. A useful general principle is to consider the factors influencing a distribution before attempts are made at transformations. Krumbein and Graybill (1965) present a table of different types of geological population distributions.

Now that the data have been established as having a normal distribution, the standard deviation and thus errors can be calculated. A quick method for calculating the standard deviation can be used with grouped data and is illustrated below using the penetrometer results.

Depth range (cm)	Mid-point of range, x_i	Frequency, f_i	$f_i x_i$	$f_i x_i^2$
31·00–31·49	31·25	1	31·25	976·56
31·50–31·99	31·75	4	127·00	4032·25
32·00–32·49	32·25	8	248·00	8320·50
32·50–32·99	32·75	10	327·50	10725·63
33·00–33·49	33·25	21	698·25	23216·81
33·50–33·99	33·75	26	877·50	29615·63
34·00–34·49	34·25	12	411·00	14076·75
34·50–34·99	34·75	5	173·75	6037·81
35·00–35·49	35·25	1	35·25	1242·56
Totals		88	2939·50	98244·50

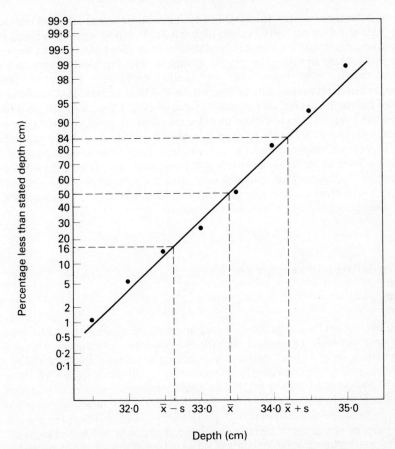

Fig. 8.7 Modified penetrometer data plotted on probability paper. The mean and one standard deviation above and below the mean are shown.

$$n = \sum f_i = 88$$

$$\bar{x} = \frac{\sum f_i x_i}{n} = \frac{2939 \cdot 50}{88} = 33 \cdot 40$$

$$s^2 = \frac{\sum f_i x_i^2}{n} - \left(\frac{\sum f_i x_i}{n}\right)^2 \tag{8.5}$$

$$= 1116 \cdot 415 - 1115 \cdot 788$$

$$= 0 \cdot 627$$

Thus

$$s = 0 \cdot 792$$

The mean of 33·40 cm can also be obtained by noting from figure 8·7 where the horizontal line through 50% intersects the best fit line to the scatter of points. The mean is then read off the *X*-axis scale. The standard deviation corresponds to the difference between the 50th and 16th percentile values and also between the 50th and 84th percentile values (figure 8.7). As can be seen from figure 8.7 the computed value for the standard deviation is about the same as the one which could be obtained by graphical means. The practical significance of these results is in being able to ascribe probability values to other penetrometer measurements falling within specified ranges. For example as noted before the probability of a value falling between $\bar{x}-s$ and $\bar{x}+s$ is 0·683. Figure 8.7 can be used to obtain any desired range. In practice results are not always plotted as in figure 8.7 and in order to obtain probability values for number ranges, appropriate tables have to be consulted. Brief consideration of figure 8.7 will suggest that the form of the distribution is dependent upon the nature of the X-axis. Thus tables of area under normal curves are only possible if the X-axis scale is converted into some standardized form. This is achieved by computing z scores which are defined as

$$z = \frac{x_i - \bar{x}}{s} \tag{8.6}$$

when dealing with sample data, or

$$z = \frac{x_i - \mu}{\sigma} \tag{8.7}$$

with population data. The relevant area under the normal curve is obtained by entering the table (Appendix 4) with the appropriate value of z and the cumulative area is thus obtained. For example, suppose the probability of obtaining penetrometer depths greater than 35·00 cm is required, then $z = (35·00 - 33·40)/0·792 = 2·020$. The cumulative area under the curve with a z-value equal to 2·020 is 0·978 and thus the probability of a value of x being greater than 35·00 cm is $(1·000 - 0·978)$ which is 0·022. In effect the z score is the number of standard deviations a value is above or below the mean.

Having introduced the concepts of standard deviation, variance, probability and the normal distribution, attention can again be focused on errors. Consider the penetrometer measurements within one square metre. A mean was obtained for 88 measurements and it is necessary to know how much confidence can be placed on this result as an estimate of the true resistance to penetration for the whole square metre. It is possible to imagine a whole series of measurements being made and for each set a mean and standard deviation can be calculated; these will be different for each set of sample values. In theory if all possible samples of similar size are drawn from a population, then the sample means will be normally distributed about the population mean as long as the samples are of reasonable size—Hammond and McCullagh (1974) suggest a minimum of 30. The standard deviation of the sampling means is called the standard error of the mean ($\sigma_{\bar{x}}$) and is obtained as follows:

$$\sigma_{\bar{x}} = \frac{\sigma}{\sqrt{n}} \tag{8.8}$$

where σ is the standard deviation of the population and n the size of the sample. In practice, of course, σ is not usually known and instead the data from one set of samples have to be used. With large samples, say over 30, the standard deviation (s) is used as an estimate of the population standard deviation (σ).

Recall the penetrometer results with $\bar{x} = 33.40$ and $s = 0.792$ and $n = 88$. Thus

$$\sigma_{\bar{x}} = \frac{0.792}{\sqrt{88}} = 0.0844$$

The estimated value for $\sigma_{\bar{x}}$ is the standard error of the mean ($\bar{x}$). In other words there is a probability of 0.683 that the true mean will lie within the range of 33.40 ± 0.084. In practice a probability value of 0.95 is required, or in special cases perhaps as high as 0.99. The range with a 0.95 probability is defined by the inequality

$$\bar{x} - \frac{1.96\sigma}{\sqrt{n}} < \mu < \bar{x} + \frac{1.96\sigma}{\sqrt{n}} \qquad (8.9)$$

For the higher probability value (0.99) the value of 1.96 must be replaced by 2.58. With the case of the penetrometer results, a 0.95 probability of being correct can be ascribed to the statement that the true mean lies within the range 33.40 ± 0.17. The range is often referred to as the confidence limits and the associated probability on a percentage basis as the confidence level. Confidence limits can be determined for any confidence level by the use of a table of cumulative normal distribution (appendix 4) and this method can be followed up in Hammond and McCullagh (1974).

Modifications to this method are required if the size of sample is less than 30 since the formula given for s (equation (8.3)) is not a good estimate of σ and also the sampling distribution of small sample means is not normal. Two procedural modifications are necessary. A better estimate of the standard deviation is achieved if a correction known as *Bessel's correction* is introduced. The resultant best estimate of σ is represented by $\hat{\sigma}$ and is defined as follows:

$$\hat{\sigma} = s\sqrt{\frac{n}{n-1}} \qquad (8.10)$$

The method of calculating this best estimate of the population standard deviation is either to compute the standard deviation using equation (8.3) and then to multiply the result by $\sqrt{n/(n-1)}$, or $\hat{\sigma}$ can be computed in one step if the following formula is used:

$$\hat{\sigma} = \sqrt{\left(\sum_{i=1}^{n}(x_i - \bar{x})^2/n - 1\right)} \qquad (8.11)$$

Elementary algebra shows that

$$\sqrt{\left(\sum_{i=1}^{n}(x_i - \bar{x})^2/n\right)} \cdot \sqrt{\frac{n}{n-1}} = \sqrt{\left(\frac{\sum_{i=1}^{n}(x_i - \bar{x})^2}{n-1}\right)}$$

Care should be taken to note that some statistical texts advocate that equation (8.11) should always be used for computing standard deviations when dealing with sample data (Dixon and Massey 1957). When the sample size is greater than 30, there is very little difference in results when equations (8.3) and (8.11) are used. However, for small samples ($n < 30$), equation (8.11) should be used. This means that the estimate of population standard deviation for the clay data can be improved when the Bessel correction is applied. Equation (8.3) gave an answer for the standard deviation of 0·161 whilst the corrected value is 0·167.

The standard error of the mean can be calculated by dividing $\hat{\sigma}$ by $\sqrt{n}$; if equations (8.8) and (8.10) are used, it can be calculated as follows:

$$\sigma_{\bar{x}} = \frac{\sigma}{\sqrt{n}} \simeq \frac{\hat{\sigma}}{\sqrt{n}} = \frac{s\sqrt{n/(n-1)}}{\sqrt{n}} = \frac{s}{\sqrt{n-1}} \tag{8.12}$$

The other modification required when dealing with samples less than 30 is to use a distribution known as '*Student's t distribution*'. Let this fact be accepted since all that need be known are two values—the confidence level and the number of degrees of freedom. As already stated 95% or 99% are the most commonly used values; again much statistical theory is omitted in order to say that the degree of freedom (df) is the number in a sample minus one for simple situations. For the clay results, the table (appendix 5) is entered at p equal to 0·05 (95% level) and df equal to 14 (15 samples minus 1). The determined value is 2·15 and this is then multiplied by the value 0·043 to give 0·092. Thus it can be postulated that the true value is 21·38 ± 0·092 with a 95% chance of being correct. This operation can be expressed more formally as:

$$\mu = \bar{x} \pm t_p \sigma_{\bar{x}} \tag{8.13}$$

where μ is the population mean, t_p is the value of Student's t statistic with $(n-1)$ degrees of freedom and p the probability, and $\sigma_{\bar{x}}$ is the standard error of the mean.

The conclusion from such a result may be that it is not sufficiently precise and thus more sample data must be obtained either to reduce the range or to increase the probability of being correct. The problem is how to calculate the required sample size for specified value ranges and confidence levels. The method can be illustrated by working through an example.

With certain geomorphological problems it is useful to assess the flatness of individual stones. One simple index of flatness can be obtained if the length (l), width (w) and thickness (t) are measured for each stone; then an index of flatness (I) can be computed as follows:

$$I = \frac{(l+w)}{2t}$$

(an index proposed by Cailleux and Tricart, quoted in King 1966). Suppose a beach is being investigated and pebbles are sampled in order to obtain flatness measures. The results of 10 randomly selected pebbles are as follows: 3·39, 2·56, 1·93, 1·29, 1·62, 1·61, 1·64, 3·22, 2·76, 1·60. Thus

$$n = 10$$
$$\bar{x} = 2·16 \quad \text{(from (8.1))}$$
$$\hat{\sigma} = 0·756 \quad \text{(from (8.11))}$$
$$\sigma_{\bar{x}} = 0·24 \quad \text{(from (8.12))}$$
$$t = 2·26 \text{ for 9 df at the 95\% level}$$

Thus

$$\mu = 2·16 \pm (2·26)(0·24) \quad \text{(from (8.13))}$$
$$= 2.16 \pm 0·54$$

This range from 1·62 to 2·70 may be considered too great and the need is to estimate the sample size required so that the range is $\pm 0·2$.

Let

$$D = t_p \sigma_{\bar{x}}$$

Then

$$D^2 = t_p^2 \sigma_{\bar{x}}^2$$

$$= t_p^2 \left(\frac{\hat{\sigma}}{\sqrt{n}} \right)^2$$

$$= t_p^2 \frac{\hat{\sigma}^2}{n}$$

$$\therefore n = \frac{t_p^2 \hat{\sigma}^2}{D^2} \tag{8.14}$$

By accepting 95% confidence limits and by taking $\hat{\sigma}$ as an estimate of the population standard deviation

$$n = \frac{(2·26)^2 (0·756)^2}{(0·2)^2}$$

$$= 73·00$$

Hence the required sample size is estimated to be 73.

This example illustrates how a decision on sample size can be made once a pilot survey has been carried out. The same general type of procedure must be adopted with more complex sampling designs, but other formulae for calculating optimum sample size have to be employed. Also it is important to stress that the preceding discussion has been concerned with data obtained by *measurement*. Other statistical techniques must be used when data have been collected through *counting*. Then the binomial distribution must be used and such an approach is described in Hammond and McCullagh (1974).

8.3 Calculation of errors in formulae

The consideration of errors has been limited to a brief description of their types followed by an outline of statistical techniques for analysing random uncertainty which is only possible if the same measurement is made several times. In practice, of course, there are many occasions when one determination only is made; often checking procedures are incorporated into the method, but the result is one measure to which may be ascribed certain confidence limits based on qualitative evidence. For example in making measurements of stone length the impression may be that these results are correct to within 2 mm. This view is usually based on the practical experience gained from making the measurements. Thus in the consideration of the index of flatness for one pebble, the length, width and thickness measurements could each be out as much as 2 mm. This statement can be made on the basis of statistical analysis of many measurements or as a result of practical experience. For the moment, let the latter approach be pursued. The index of flatness also illustrates the

F

point that measurements are often combined in some form in order to produce a measure and it is thus important to be able to assess the magnitude of the error of this value.

Before this problem is tackled, some notation is necessary to describe errors, and the terms as used by Topping (1972) are adopted. Suppose an unchanging feature has a true measurement of x_0 units, then x_0 can be estimated by taking several measurements and calculating the mean, i.e.

$$\bar{x} = \frac{x_1 + x_2 + x_3 + \ldots + x_n}{n}$$

where $\bar{x}$ is the mean and x_1, x_2, etc. are the individual values which total n in number. The difference between the true value (x_0) and the best estimate ($\bar{x}$) is explained by the magnitude of the total error (e_r). Thus

$$\bar{x} = x_0 + e_r \tag{8.15}$$

and e_r can have a positive or negative value. It is also possible to express equation (8.15) as

$$\bar{x} = x_0(1 + f) \tag{8.16}$$

if f equals e_r/x_0; f is called the fractional error in x_0 and can also be expressed as a percentage if multiplied by 100. It is important to stress that the value of e_r in equation (8.15) is the total *actual* error rather than the maximum limit. For example, a length can be measured to ± 2 mm, but for a particular measurement the actual error (e) may be $+0.5$ mm. The limiting error may be represented as E and any actual error would be less than this. A distinction needs to be made between e_r and E when errors associated with values which are derived from several measurements are estimated.

Error in a sum or difference

The index of flatness involved the addition of two measurements and the result was divided by another in order to yield a value for the index (I). To tackle the error in I, the first concern is to estimate the error resultant upon the addition of two length measures. Such a task is deceptively simple. Let some actual values for adding a length and width be used: suppose the length (l) of one stone is 77 mm, the width (w) is 45 mm and these values are correct to within 2 mm. Thus ($l + w$) could vary from 118 to 126 mm, but it is very unlikely that these extreme values would be encountered. This can be explained by brief consideration of probability using the notation as described above. Suppose

$$e_1 \geq E_1 \text{ with } p_1 = 0.05$$

and

$$e_2 \geq E_2 \text{ with } p_2 = 0.05$$

then for

$$(e_1 + e_2) \geq (E_1 + E_2), \ p = p_1 p_2 = 0.0025$$

This means that it is justifiable to search for a 'most probable' limiting error value which lies between the individual limits, E_1 and E_2, and their sum ($E_1 + E_2$). The practice is to take the most probable value as $\sqrt{(E_1^2 + E_2^2)}$. Thus the most probable maximum error resulting upon adding the length measures is

$\sqrt{(2^2 + 2^2)}$ which is 2·828. The most probable limiting value for the fractional error can now be obtained by dividing 2·828 by $(77 + 45)$ to give 0·023. Exactly the same result would be obtained had it been necessary to *subtract* 45 from 77 mm.

Error in a product and in a quotient

When a value is obtained by multiplying several measurements, then the overall fractional error can be estimated by adding the constituent fractional errors. If, for example, $Q = abc$, the fractional error in Q can be calculated by adding the fractional errors of a, b and c. Similarly if $Q = a/b$, then the fractional error in Q is approximated by subtracting the fractional error of b from that of a (Topping 1972). Multiplication or division by a constant value does not change the value of a fractional error.

More complex formulae

These algebraic procedures can be followed for simple formulae when the overall error has to be assessed, but resort needs to be made to calculus for more complex problems. This approach can be demonstrated by considering the situation whereby the quantity y is predicted from x and $y = f(x)$. The error in x can be represented by δx and the error in y is similarly δy. By definition,

$$\lim_{\delta x \to 0} \frac{\delta y}{\delta x} = \frac{\mathrm{d}y}{\mathrm{d}x},$$

and if δx is very small, $\delta y/\delta x \simeq \mathrm{d}y/\mathrm{d}x$. Thus the error in y can be estimated if $(\mathrm{d}y/\mathrm{d}x)\delta x$ is determined. Suppose x is the radius of a sphere and y the volume, then $y = \frac{4}{3}\pi x^3$ and $\mathrm{d}y/\mathrm{d}x = 4\pi x^2$

Hence

$$\frac{\delta y}{\delta x} \simeq 4\pi x^2 \text{ if } \delta x \text{ is small}$$

Thus

$$\delta y = 4\pi x^2 \delta x$$

The fractional error in the volume determination is

$$\frac{\delta y}{y} = \frac{4\pi x^2 \delta x}{\frac{4}{3}\pi x^3} = \frac{3\delta x}{x}$$

This means that the fractional error in the volume is three times that of the fractional error in the radius.

The use of calculus comes into its own when a quantity is a function of several variables. If $Q = f(x, y, z, \ldots)$, the error in Q (δQ) results from the constituent errors (δx, δy, δz ...,) and

$$\delta Q \simeq \frac{\partial Q}{\partial x}\delta x + \frac{\partial Q}{\partial y}\delta y + \frac{\partial Q}{\partial z}\delta z + \ldots \tag{8.17}$$

It should be explained that the term $(\partial Q/\partial x)\delta x$ is a partial differential which expresses the partial error in Q because of the error δx in x only; the other partial differentials can be similarly described.

A simple example may help with comprehension. Suppose it is possible to predict stream discharge (Q) at a point by making a measurement of stream velocity (v), width (w) and depth (d). Then

$$Q = vwd$$

From equation (8.17)

$$Q \simeq \frac{\partial Q}{\partial v}\delta v + \frac{\partial Q}{\partial w}\delta w + \frac{\partial Q}{\partial d}\delta d$$

$$\simeq wd\delta v + vd\delta w + vw\delta d$$

$$\therefore \quad \frac{\delta Q}{Q} \simeq \frac{\delta v}{v} + \frac{\delta w}{w} + \frac{\delta d}{d}$$

This establishes an already stated fact, namely that for a product, the overall fractional error is obtained by adding the component fractional errors. The use of partial derivatives becomes essential when dealing with such formulae as

$$y = \frac{pqr^3 s}{tuv}$$

where p, q, r, s, t, u and v are variables.

A useful preliminary step is to take logarithms:

$$\log y = \log\frac{pqr^3 s}{tuv}$$

$$= \log pqr^3 s - \log tuv$$

$$= \log p + \log q + 3 \log r + \log s - \log t - \log u - \log v$$

$$\therefore \quad \frac{\delta y}{y} \simeq \frac{\delta p}{p} + \frac{\delta q}{q} + \frac{3\delta r}{r} + \frac{\delta s}{s} - \frac{\delta t}{t} - \frac{\delta u}{u} - \frac{\delta v}{v}$$

The important term is $3\delta r/r$ since its value can have a much greater effect on $\delta y/y$ than the other terms; this situation would be particularly marked if r happened to have a small value. The assumption in equation (8.17) is that the component maximum errors may *all* occur for one determination of Q; this is possible, but as discussed earlier it is extremely unlikely. Thus it is more realistic to use another equation which estimates the 'most probable' maximum value of δQ:

$$(\delta Q)^2 \simeq \left(\frac{\partial Q}{\partial x}\delta x\right)^2 + \left(\frac{\partial Q}{\partial y}\delta y\right)^2 + \left(\frac{\partial Q}{\partial z}\delta z\right)^2 + \dots \qquad (8.18)$$

The results from equations (8.17) and (8.18) can be compared for the stream discharge example: $Q = vwd$. Suppose

$$v = 0{\cdot}50 \pm 0{\cdot}05 \text{ m s}^{-1}$$
$$w = 5{\cdot}0 \pm 0{\cdot}01 \text{ m}$$
$$d = 1{\cdot}0 \pm 0{\cdot}01 \text{ m}$$

From equation (8.17)

$$\delta Q \simeq \frac{\partial Q}{\partial v}\delta v + \frac{\partial Q}{\partial w}\delta w + \frac{\partial Q}{\partial d}\delta d$$

$$\simeq wd\delta v + vd\delta w + vw\delta d$$

$$= 5\cdot0 \times 1\cdot0 \times 0\cdot05 + 0\cdot50 \times 1\cdot0 \times 0\cdot01 + 0\cdot50 \times 5\cdot0 \times 0\cdot01$$

$$= 0\cdot28$$

From equation (8.18)

$$(\delta Q)^2 = \left(\frac{\partial Q}{\partial v}\delta v\right)^2 + \left(\frac{\partial Q}{\partial w}\delta w\right)^2 + \left(\frac{\partial Q}{\partial d}\delta d\right)^2$$

$$\simeq (wd\delta v)^2 + (vd\delta w)^2 + (vw\delta d)^2$$

$$= (5\cdot0 \times 1\cdot0 \times 0\cdot05)^2 + (0\cdot50 \times 1\cdot0 \times 0\cdot01)^2 + (0\cdot50 \times 5\cdot0 \times 0\cdot01)^2$$

$$= 0\cdot06315$$

$$\therefore \quad \delta Q = 0\cdot25$$

These results can be compared by saying that there is almost complete certainty that Q lies within $\pm 0\cdot28$ of the mean, but for most practical purposes a prediction that $Q = 2\cdot50 \pm 0\cdot25$ m^3 s^{-1} is quite satisfactory.

The assumption in the preceding discussion is that for any measure an estimate of the maximum associated error can be given. Estimates of errors can be made if the same measurement is repeated several times so that the standard error of the mean is calculated. When a new value is calculated from several other measurements, each with specific standard errors of their means, similar procedures can be followed to those already described in order to estimate the standard error of the new value. Suppose that V is a value which is computed from measurements of $a, b, c, \ldots$ and suppose that the latter have $\sigma_{\bar{a}}, \sigma_{\bar{b}}, \sigma_{\bar{c}}, \ldots$ as standard errors of their means. Then the problem is to predict the standard error of the mean for V represented by $\sigma_{\bar{V}}$ if V is dependent upon different types of algebraic relationships, $\bar{a}$, $\bar{b}$ and $\bar{c}$ represent the means of these variables.

(a) sum or difference

$$V = a \pm b \tag{8.19}$$

$$\sigma_{\bar{V}} = \pm \sqrt{\sigma_{\bar{a}}^2 + \sigma_{\bar{b}}^2} \tag{8.20}$$

(b) Product

$$V = abc \tag{8.21}$$

$$\frac{\sigma_{\bar{V}}}{V} = \pm \sqrt{\left(\frac{\sigma_{\bar{a}}^2}{\bar{a}^2} + \frac{\sigma_{\bar{b}}^2}{\bar{b}^2} + \frac{\sigma_{\bar{c}}^2}{\bar{c}^2}\right)} \tag{8.22}$$

(c) Quotient

$$V = a/b \tag{8.23}$$

$$\frac{\sigma_{\bar{V}}}{V} = \pm \sqrt{\left(\frac{\sigma_{\bar{a}}^2}{\bar{a}^2} + \frac{\sigma_{\bar{b}}^2}{\bar{b}^2}\right)} \tag{8.24}$$

(d) General case

$$V = a^x b^y c^z \tag{8.25}$$

$$\frac{\sigma_{\bar{V}}}{V} = \pm \sqrt{\left(\frac{x^2 \sigma_{\bar{a}}^2}{\bar{a}^2} + \frac{y^2 \sigma_{\bar{b}}^2}{\bar{b}^2} + \frac{z^2 \sigma_{\bar{c}}^2}{\bar{c}^2} \right)} \tag{8.26}$$

(e) Constant multiple

$$V = xa \text{ where } x \text{ is a constant} \tag{8.27}$$

$$\frac{\sigma_{\bar{V}}}{V} = \pm \frac{\sigma_{\bar{a}}}{\bar{a}} \tag{8.28}$$

(after Whittle and Yarwood 1973).

A large number of these formulae are possible; their derivations are dependent upon calculus as previously demonstrated. Certain general observations can be made from these formulae about errors. With a sum or difference the largest error will dominate the result, and far more so when powers are involved. Thus the power of a variable has a tremendous influence on the resultant overall error. Another conclusion from these formulae is that measurement techniques for one experiment need to be coordinated since it is pointless to include one very time-consuming precise measurement if others are to be made with far less precision in the same experiment (Braddick 1963).

8.4 Reporting of results

The culmination of experimental work is the reporting of results and the aim of measurement is to produce results with an indication of errors. Before attention is focused on errors, a comment on the presentation of numbers is appropriate. The decimal sign between digits is usually a point, but an alternative is a comma. For large numbers, the digits may be grouped in threes about the decimal point, for example

6 354·189 432 but not 6,354·189,432

Also, when the decimal point precedes the first digit of a number, a zero should be inserted before the decimal sign, for example

0·438 but not ·438

Associated with each determination there should be not only degrees of precision, but also degrees of confidence. For example, the average altitude of a planation surface could be reported as 163 ± 9 m ($p = 0.95$). In other words this statement is saying that within a particular area there is a 95% chance of the altitude of the surface being more than 154 m and less than 172 m O.D. If geomorphologists had reported their results in this form in the past, a great deal of the controversy over the correlation of surfaces would not have arisen. When many measurements are made of the same phenomenon, confidence limits and levels can be quoted, but this is not possible when single measurements are made, for example the temperature at a locality at a specific time. But even in such a situation some indication of precision can be given by the number of significant digits recorded for the temperature. An answer of 20 °C implies 20 ± 0.5 °C whilst 20·1 °C indicates a range of 20.1 ± 0.05 °C.

Thus it is essential to report a result consistent with the error associated with the measurement. Care must be taken when further calculations are made that final results do not imply a higher level of precision than is justified. As a rule of thumb when determinations are multiplied or divided, the number of significant figures in the final result cannot exceed that in the least precise measurement. Suppose a density calculation is to be made from the following results:

$$\text{mass} \quad 13\cdot718 \text{ g}$$
$$\text{volume} \quad 13 \text{ cm}^3$$

There are only two significant figures in the 13 cm^3 measurement and therefore the answer has to be reported as $1\cdot1$ g cm^{-3} or 11×10^2 kg m^{-3}. However it must be stressed that the results from such computations need not have the same precision as the initial measurements, and indeed further errors can be introduced. To exemplify this point, Bragg (1974) selects π whose two-digit value is $3\cdot1$, that is π lies in the range $3\cdot05$ to $3\cdot15$. π^2 has a value of $9\cdot61$ which has to be shortened to $9\cdot6$ following the above rule. In fact π^2 has a value of $9\cdot8696 \dots$. Erroneous answers are also obtained if three or more digits are selected for π. The important point to note is that rounding of an answer to the same number of digits as the minimum number of an included measurement can lead to misleading results.

Suppose seven measurements were made and their numerical values were reported as follows: $4\cdot6$, $4\cdot4$, $5\cdot1$, $4\cdot8$, $3\cdot9$, $4\cdot5$ and $4\cdot7$. The implication is that these values have maximum errors of the order of $\pm0\cdot05$. The sum of these values is $32\cdot0$ but if all the seven values happened to have their maximum or minimum error, then the sum would be $32\cdot35$ or $31\cdot65$ respectively. How should the arithmetic mean of these seven values be reported? If their sum is divided by seven, the result is $4\cdot571\,429$ to six decimal places, but such a statement implies an error of $\pm0\cdot000\,000\,5$ and clearly would be misleading. From the maximum and minimum sums the value of the average could range from $4\cdot621\,429$ to $4\cdot521\,429$. If the average was reported as $4\cdot57$, the implied range of $4\cdot565$ to $4\cdot575$ would only be a part of the possible range. Instead, the mean could be reported as $4\cdot6$ with the associated maximum range of $4\cdot55$ to $4\cdot65$. This statement is not entirely satisfactory either since the range $4\cdot53$ to $4\cdot56$ is omitted and values above $4\cdot62$ are virtually impossible. An alternative approach is to calculate the most probable maximum error (E). This is determined in this case by adding the squares of the seven individual errors $(0\cdot05)$ and taking the square root—the result is $0\cdot132$, which must then be divided by 7 to give $0\cdot018\,9$. Thus it can be said with reasonable confidence that the mean of the seven values lies within $0\cdot018\,9$ of $4\cdot571\,429$. The implication is that the mean, if reported as 4.57, indicates a smaller associated error range $(4\cdot565$ to $4\cdot575)$ than that obtained from the maximum probable error range $(4\cdot552\,5$ to $4\cdot590\,3)$. Again, the reporting of the mean to one decimal place seems the best action. Of course it is possible to analyse the error problem in greater detail by computing the standard error of the mean, but the example illustrates the point that errors have to be considered even when the simple operation of determining an arithmetic mean is undertaken.

Every reader is without doubt familiar with rounding of numbers, but several suggestions made by Campion *et al.* (1973) are relevant to this consideration of results. They propose that if a decimal place with a 5 in it is to be discarded, then the preceding digit should be changed only if it is an odd number. For example, 3·845 can be rounded to 3·84 whilst 3·855 would change to 3·86. Such a procedure does not seem to be widely adopted. They stress that rounding should always be done in one operation so that 2·346 rounds to 2·3 rather than 2·4 if two steps had been taken. With errors they propose that rounding should always be upwards to give a more pessimistic estimate of precision.

The aim of any experimental programme is the production of the results and the form of the presentation will depend on the nature of the problem under investigation. The ultimate form of result statement could be as follows:

Mean bulk density	1253 kg m^{-3}
Standard error of mean (df = 10)	5 kg m^{-3}
Estimate of total systematic uncertainty	40 kg m^{-3}

There is little need to point out that it is very rare for physical geographers to present their results in such a refined form. The criticism has already been made that there has been a tendency to omit error considerations. In part this can be explained by the fact that the usual concern is to establish patterns in space or time which can then be interpreted. In contrast it is very much the task of the physicist or chemist to determine constant values. Nevertheless, trends or patterns in results from physical geography can only be convincingly established if the analysis includes consideration of errors. For example, in process geomorphology, the assessment of errors is particularly relevant. A process such as soil creep is difficult to monitor given its very slow rate. There are, of course, many technical difficulties associated with the measurement of this process. Most experiments only manage to obtain data on creep for a few years, but any suggestion about how much soil has moved say in 1000 years must only be made after careful assessment of errors in the original data.

The final phase of scientific investigation, namely the interpretation of the results, can only begin after the results are obtained. It is this interpretation which permits a judgement to be made on the problem as originally defined. The experimental results are the means to this end, and this re-iterates the theme that decisions on such matters as required precision can only be made with reference to the problem under investigation. Physical geographers or environmental scientists need to be far more conscious of the necessity for rigorous statements of problems. This is not to deny the very important role of description in the subject, but such exercises as geomorphological, vegetation or soil mapping should be seen as steps towards the identification of specific problems for investigation. Often the formulation and investigation of a problem requires sound comprehension of physics or chemistry, and the derivation and ultimate problem statement as well as the analysis may use mathematics. With the attainment of such scientific expertise, the contribution of physical geographers to an understanding of the physical environment should be increasingly evident.

APPENDIX 1

Examples of SI derived units including those with special names (based on Royal Society 1975).

Quantity	Name	Expression in terms of other units	Expression in terms of SI base units
area	square metre		m^2
volume	cubic metre		m^3
velocity	metre per second		$m\ s^{-1}$
acceleration	metre per second squared		$m\ s^{-2}$
density	kilogram per cubic metre		$kg\ m^{-3}$
concentration	mole per cubic metre		$mol\ m^{-3}$
force	newton (N)		$m\ kg\ s^{-2}$
pressure	pascal (Pa)	$N\ m^{-2}$	$m^{-1}\ kg\ s^{-2}$
energy, work, quantity of heat	joule (J)	$N\ m$	$m^2\ kg\ s^{-2}$
power, radiant flux	watt (W)	$J\ s^{-1}$	$m^2\ kg\ s^{-3}$
dynamic viscosity	pascal second	$Pa\ s$	$m^{-1}\ kg\ s^{-1}$
kinematic viscosity	square metre per second		$m^2\ s^{-1}$
moment of force	newton metre	$N\ m$	$m^2\ kg\ s^{-2}$
surface tension	newton per metre	$N\ m^{-1}$	$kg\ s^{-2}$
heat capacity	joule per kelvin	$J\ K^{-1}$	$m^2\ kg\ s^{-2}\ K^{-1}$
thermal conductivity	watt per metre kelvin	$W\ m^{-1}\ K^{-1}$	$m\ kg\ s^{-3}\ K^{-1}$

APPENDIX 2

Conversion of various units to equivalent values in SI system

Length:

1 mile	$= 1\cdot609\,34$ km
1 yard	$= 0\cdot914\,4$ m
1 foot	$= 0\cdot304\,8$ m
1 inch	$= 25\cdot4$ mm
1 fathom	$= 1\cdot828\,8$ m
1 Å (ångström)	$= 10^{-10}$ m

Area:

1 square mile	$= 2\cdot589\,99$ km^2 $= 258\cdot999$ ha
1 acre	$= 4046\cdot86$ m^2 $= 0\cdot404\,686$ ha
1 square yard	$= 0\cdot836\,127$ m^2
1 square foot	$= 0\cdot092\,903\,0$ m^2
1 square inch	$= 645\cdot16$ mm^2
1 ha (hectare)	$= 10^4$ m^2

Volume:

1 cubic yard	$= 0\cdot764\,555$ m^3
1 cubic foot	$= 28\cdot316\,8$ dm^3
1 cubic inch	$= 16\cdot387\,1$ cm^3

Velocity:

1 mile per hour	$= 1\cdot609\,34$ km h^{-1}
1 foot per second	$= 0\cdot304\,8$ m s^{-1}

Acceleration:

1 foot per second per second	$= 0\cdot304\,8$ m s^{-2}

Mass:

1 ton	$= 1016\cdot05$ kg $= 1\cdot016\,05$ t
1 pound	$= 0\cdot453\,592\,37$ kg
1 t (metric ton)	$= 1000$ kg

Volume rate of flow:

1 cubic foot per second	$= 0\cdot028\,316\,8$ m^3 s^{-1}

Force:

1 dyne	$= 10^{-5}$ N

Pressure, stress:
1 atmosphere	$= 101\cdot325 \text{ kN m}^{-2}$
1 bar	$= 10^5 \text{ N m}^{-2}$
1 millibar	$= 100 \text{ N m}^{-2}$
1 mm mercury	$= 133\cdot322 \text{ N m}^{-2}$
1 mm water	$= 9\cdot806\,65 \text{ N m}^{-2}$
1 pound force per square inch	$= 6\cdot894\,76 \text{ kN m}^{-2}$

Other useful pressure conversions:
1 atmosphere	$= 1\cdot013\,25 \text{ bar}$
	$= 76 \text{ cm mercury}$
	$= 10\cdot332\,6 \text{ m water}$
1 bar	$= 9\cdot869\,23 \times 10^{-1} \text{ atmosphere}$
	$= 75\cdot006\,2 \text{ cm mercury}$
1 cm mercury	$= 1\cdot315\,789\,5 \times 10^{-2} \text{ atmosphere}$
	$= 1\cdot333\,22 \times 10^{-2} \text{ bar}$
1 cm water	$= 9\cdot678\,14 \times 10^{-4} \text{ atmosphere}$

Energy, work, heat: (see note 1 about the calorie)
1 therm	$= 105\cdot506 \text{ MJ}$
1 kilowatt hour	$= 3\cdot6 \text{ MJ}$
1 British thermal unit	$= 1\cdot055\,06 \text{ kJ}$

Power:
1 horse power	$= 0\cdot745\,700 \text{ kW}$

Temperature:
Procedure for converting from
   (1) Fahrenheit to Celsius
       Subtract 32 from the Fahrenheit value and then take 5/9ths of the result
       to obtain the Celsius value.
   (2) Fahrenheit to Kelvin
       Add 459·67 to the Fahrenheit value and then take 5/9ths of the result to
       obtain the Kelvin value.

Note 1:
The calorie and joule
The joule is the unit of energy in the SI system. It is the work done when the point of application of a force of one newton is moved through one metre in the direction of the force.

    This joule is occasionally called the absolute joule to distinguish it from the international joule, a unit which should be no longer used.

    1 international joule $= 1\cdot000\,19$ absolute joule.

    A widely used unit of heat is the calorie, but it is gradually being abandoned since it does not conform to the SI system. A calorie is the amount of heat necessary to raise the temperature of 1 g water at 15 °C by 1 C°. 1000 calories or 1 kilocalorie is sometimes called 1 Calorie, a unit formerly used, for example, in the heat equivalent of foods. In detail, three types of calorie are in use: (a) the calorie as defined above by reference to raising 1 g of water at 15 °C by 1 C°— labelled cal_{15}; (b) the international calorie as defined by an International

Steam Table Conference—labelled cal_{IT}; (c) the thermochemical calorie—labelled cal_{therm}.

Conversion of these calories to absolute joules (J):

$$1\ cal_{15} = 4\cdot1855\ J$$
$$1\ cal_{IT} = 4\cdot1868\ J$$
$$1\ cal_{therm} = 4\cdot184\ J$$

Also

$$1\ cal_{IT} = 4\cdot1860\ \text{international joules.}$$

It is clear that errors can easily arise in the conversion of calories to joules if the exact type of calories and joules are not specified. In most chemistry texts concern is with thermochemical calories whilst physics texts use the other two calories. The only consolation is that the differences in the conversion factors are so small that any resultant errors can, for most practical purposes, be ignored.

Sources for conversions:
Anderton and Bigg (1967)
Parrish (1969)
Weast (1974)

Appendix 3

The Periodic Table

I A	II A	III B	IV B	V B	VI B	VII B
1 **H** 1·0080						
3 **Li** 6·941	4 **Be** 9·012					
11 **Na** 22·99	12 **Mg** 24·31					

		III B	IV B	V B	VI B	VII B	
19 **K** 39·102	20 **Ca** 40·08	21 **Sc** 44·96	22 **Ti** 47·90	23 **V** 50·94	24 **Cr** 52·00	25 **Mn** 54·94	26 **Fe** 55·85
37 **Rb** 85·47	38 **Sr** 87·62	39 **Y** 88·91	40 **Zr** 91·22	41 **Nb** 92·91	42 **Mo** 95·94	43 **Tc** (99)	44 **Ru** 101·07
55 **Cs** 132·91	56 **Ba** 137·34	57 **La*** 138·91	72 **Hf** 178·49	73 **Ta** 180·95	74 **W** 183·85	75 **Re** 186·2	76 **Os** 190·2
87 **Fr** (223)	88 **Ra** (226)	89 **Ac**† (227)	104 (260)	105			

	58	59	60	61
*Lanthanide Series	**Ce** 140·12	**Pr** 140·91	**Nd** 144·24	**Pm** (147)
†Actinide Series	90 **Th** 232·04	91 **Pa** (231)	92 **U** 238·03	93 **Np** (237)

							VIII A
							2 **He** 4·0026

		III A	IV A	V A	VI A	VII A	
		5 **B** 10·81	6 **C** 12·011	7 **N** 14·007	8 **O** 15·999	9 **F** 18·998	10 **Ne** 20·179
I B	II B	13 **Al** 26·98	14 **Si** 28·09	15 **P** 30·974	16 **S** 32·06	17 **Cl** 35·453	18 **Ar** 39·948
29 **Cu** 63·55	30 **Zn** 65·37	31 **Ga** 69·72	32 **Ge** 72·59	33 **As** 74·92	34 **Se** 78·96	35 **Br** 79·904	36 **Kr** 83·80
47 **Ag** 107·87	48 **Cd** 112·40	49 **In** 114·82	50 **Sn** 118·69	51 **Sb** 121·75	52 **Te** 127·60	53 **I** 126·90	54 **Xe** 131·30
79 **Au** 196·97	80 **Hg** 200·59	81 **Tl** 204·37	82 **Pb** 207·2	83 **Bi** 208·98	84 **Po** (210)	85 **At** (210)	86 **Rn** (222)

64 **Gd** 157·25	65 **Tb** 158·92	66 **Dy** 162·50	67 **Ho** 164·93	68 **Er** 167·26	69 **Tm** 168·93	70 **Yb** 173·04	71 **Lu** 174·97
96 **Cm** (247)	97 **Bk** (247)	98 **Cf** (251)	99 **Es** (254)	100 **Fm** (253)	101 **Md** (256)	102 **No** (254)	103 **Lw** (257)

Appendix 4

Appendix 4 Cumulative Normal Distribution

z	Area	z	Area	z	Area	z	Area
−3·25	·0006	−1·00	·1587	1·05	·8531	−4·265	·00001
−3·20	·0007	− ·95	·1711	1·10	·8643	−3·719	·0001
−3·15	·0008	− ·90	·1841	1·15	·8749	−3·090	·001
−3·10	·0010	− ·85	·1977	1·20	·8849	−2·576	·005
−3·05	·0011	− ·80	·2119	1·25	·8944	−2·326	·01
−3·00	·0013	− ·75	·2266	1·30	·9032	−2·054	·02
−2·95	·0016	− ·70	·2420	1·35	·9115	−1·960	·025
−2·90	·0019	− ·65	·2578	1·40	·9192	−1·881	·03
−2·85	·0022	− ·60	·2743	1·45	·9265	−1·751	·04
−2·80	·0026	− ·55	·2912	1·50	·9332	−1·645	·05
−2·75	·0030	− ·50	·3085	1·55	·9394	−1·555	·06
−2·70	·0035	− ·45	·3264	1·60	·9452	−1·476	·07
−2·65	·0040	− ·40	·3446	1·65	·9505	−1·405	·08
−2·60	·0047	− ·35	·3632	1·70	·9554	−1·341	·09
−2·55	·0054	− ·30	·3821	1·75	·9599	−1·282	·10
−2·50	·0062	− ·25	·4013	1·80	·9641	−1·036	·15
−2·45	·0071	− ·20	·4207	1·85	·9678	− ·842	·20
−2·40	·0082	− ·15	·4404	1·90	·9713	− ·674	·25
−2·35	·0094	− ·10	·4602	1·95	·9744	− ·524	·30
−2·30	·0107	− ·05	·4801	2·00	·9772	− ·385	·35
−2·25	·0122			2·05	·9798	− ·253	·40
−2·20	·0139			2·10	·9821	− ·126	·45
−2·15	·0158	·00	·5000	2·15	·9842	0	·50
−2·10	·0179			2·20	·9861	·126	·55
−2·05	·0202			2·25	·9878	·253	·60
−2·00	·0228	·05	·5199	2·30	·9893	·385	·65
−1·95	·0256	·10	·5398	2·35	·9906	·524	·70
−1·90	·0287	·15	·5596	2·40	·9918	·674	·75
−1·85	·0322	·20	·5793	2·45	·9929	·842	·80
−1·80	·0359	·25	·5987	2·50	·9938	1·036	·85
−1·75	·0401	·30	·6179	2·55	·9946	1·282	·90
−1·70	·0446	·35	·6368	2·60	·9953	1·341	·91
−1·65	·0495	·40	·6554	2·65	·9960	1·405	·92
−1·60	·0548	·45	·6736	2·70	·9965	1·476	·93
−1·55	·0606	·50	·6915	2·75	·9970	1·555	·94
−1·50	·0668	·55	·7088	2·80	·9974	1·645	·95
−1·45	·0735	·60	·7257	2·85	·9978	1·751	·96
−1·40	·0808	·65	·7422	2·90	·9981	1·881	·97
−1·35	·0885	·70	·7580	2·95	·9984	1·960	·975
−1·30	·0968	·75	·7734	3·00	·9987	2·054	·98
−1·25	·1056	·80	·7881	3·05	·9989	2·326	·99
−1·20	·1151	·85	·8023	3·10	·9990	2·576	·995
−1·15	·1251	·90	·8159	3·15	·9992	3·090	·999
−1·10	·1357	·95	·8289	3·20	·9993	3·719	·9999
−1·05	·1469	1·00	·8413	3·25	·9994	4·265	·99999

Note
Source: Dixon, W. J. and Massey, F. J. 1957: *Introduction to Statistical Analysis,* second edition (New York) pp. 382–383.

Appendix 5

Student's t-Distribution

Table of values of *t* corresponding to specified *two-tailed* probabilities and degrees of freedom

Degrees of freedom	p = 0·1 p'= 90%	p = 0·05 p'= 95%	p = 0·02 p'= 98%	p = 0·01 p'= 99%	p = 0·001 p'= 99·9%
1	6·31	12·71	31·82	63·66	636·62
2	2·92	4·30	6·97	9·93	31·60
3	2·35	3·18	4·54	5·84	12·94
4	2·13	3·78	3·75	4·60	8·61
5	2·02	2·57	3·37	4·03	6·86
6	1·94	2·45	3·14	3·71	5·96
7	1·90	2·37	3·00	3·50	5·41
8	1·86	2·31	2·90	3·36	5·04
9	1·83	2·26	2·82	3·25	4·78
10	1·81	2·23	2·76	3·17	4·59
11	1·80	2·20	2·72	3·11	4·44
12	1·78	2·18	2·68	3·06	4·32
13	1·77	2·16	2·65	3·01	4·22
14	1·76	2·15	2·62	2·98	4·14
15	1·75	2·13	2·60	2·95	4·07
16	1·75	2·12	2·58	2·92	4·02
17	1·74	2·11	2·57	2·90	3·97
18	1·73	2·10	2·55	2·88	3·92
19	1·73	2·09	2·54	2·86	3·88
20	1·73	2·09	2·53	2·85	3·85
21	1·72	2·08	2·52	2·83	3·82
22	1·72	2·07	2·51	2·82	3·79
23	1·71	2·07	2·50	2·81	3·77
24	1·71	2·06	2·49	2·80	3·75
25	1·71	2·06	2·49	2·79	3·73
26	1·71	2·06	2·48	2·78	3·71
27	1·70	2·05	2·47	2·77	3·69
28	1·70	2·05	2·47	2·76	3·67
29	1·70	2·05	2·46	2·76	3·66
30	1·70	2·04	2·46	2·75	3·65
40	1·68	2·02	2·42	2·70	3·55
60	1·67	2·00	2·39	2·66	3·46

Note:
p is the two-tailed probability of a value being more extreme than *t*.
p' is the two-tailed probability (expressed as a percentage) of a value being less extreme than *t*.

From Fisher and Yates: *Statistical Tables for Biological, Agricultural and Medical Research*, published by Longman Group Ltd. London, (previously published by Oliver and Boyd, Edinburgh), and by permission of the authors and publishers.

Suggestions for further reading

Introduction

Introductory texts on physics and chemistry designed for particular subject areas within physical geography or environmental science.

BYERS, H. R. 1965: *Elements of cloud physics.* Chicago: University of Chicago Press.
DUNCAN, G. 1975: *Physics for biologists.* Oxford: Blackwell.
GYMER, R. G. 1973: *Chemistry: an ecological approach.* New York: Harper and Row.
MONTEITH, J. L. 1973: *Principles of environmental physics.* London: Arnold.
ROGERS, R. R. 1976: *A short course in cloud physics.* Oxford: Pergamon Press.
STAMPER, J. G. and STAMPER, N. A. 1971: *Chemistry for biologists.* London: Allen and Unwin.

Chapter 1

The SI system:
1973: *SI the International System of Units.* Prepared jointly by the National Physical Laboratory, UK and the National Bureau of Standards, USA. (London and Washington).
1975: *Quantities, Units and Symbols.* A report by the Symbols Committee of the Royal Society. (London).
1972: *The Use of SI Units.* The British Standards Institution PD 5686.

Chapter 2

ANDERSON, C. B., FORD, P. C. and KENNEDY, J. H. 1973: *Chemistry: principles and applications.* Lexington: D. C. Heath and Co.
GYMER, R. G. 1973: *Chemistry: an ecological approach.* New York: Harper and Row.
KEENAN, C. W., WOOD, J. H. and KLEINFELTER, D. C. 1976: *General college chemistry,* fifth edition. New York: Harper and Row.
MASTERTON, W. L. and SLOWINSKI, E. J. 1973: *Chemical principles.* Philadelphia: Saunders.
PAULING, L. 1970: *General chemistry,* third edition. San Francisco: Freeman.
PIERCE, J. B. 1970: *The chemistry of matter.* Boston: Houghton Mifflin.
SIENKO, M. J. and PLANE, R. A. 1974: *Chemical principles and properties,* second edition. New York: McGraw-Hill.
1976: *Chemistry,* fifth edition. New York: McGraw-Hill.

Chapter 3

HALLIDAY, D. and RESNICK, R. 1966: *Physics,* combined edition. New York: Wiley.
SEARS, F. W. and ZEMANSKY, M. W. 1963: *University physics,* third edition. Reading, Mass.: Addison-Wesley.
SEARS, F. W., ZEMANSKY, M. W. and YOUNG, H. D. 1974: *College physics,* fourth edition. Reading, Mass.: Addison-Wesley.

TILLEY, D. and THUMM, W. 1971: *College physics with applications to the life sciences.* Menlo Park, California: Cummings Publishing Co.

Chapter 4

The texts as listed for chapter 2 are recommended.

Chapter 5

The chemistry texts as listed for chapter 2 are recommended.
In addition:
CURTIS, C. D., 1976: Chemistry of rock weathering: fundamental reactions and controls. In Derbyshire, E., editor, *Geomorphology and climate,* 25–57, London: Wiley.

Chapter 6

The texts on physics as listed for chapter 3 are recommended.
In addition:
MONTEITH, J. L. 1973: *Principles of environmental physics.* London: Arnold.

Chapter 7

The chemistry and physics texts as listed for chapters 2 and 3 are recommended.
In addition:
CAPPER, P. L. and CASSIE, W. F. 1963: *The mechanics of engineering soils,* fourth edition. London: Spon.
CARSON, M. A. 1971: *The mechanics of erosion.* London: Pion.
JUMIKIS, A. R. 1967: *Introduction to soil mechanics.* Princetown, N.J.: Van Nostrand.
MASSEY, B. S. 1975: *Mechanics of fluids,* third edition. New York: Van Nostrand.
STREETER, V. L. and WYLIE, E. B. 1975: *Fluid mechanics,* sixth edition. New York: McGraw-Hill.
TERZAGHI, K. 1943: *Theoretical soil mechanics.* New York: Wiley.
VENNARD, J. K. and STREET, R. L. 1975: *Elementary fluid mechanics,* fifth edition. New York: Wiley.

Chapter 8

ACKOFF, R. L. 1962: *Scientific method in optimising applied research decisions.* New York: Wiley.
BRAGG, G. M. 1974: *Principles of experimentation and measurement.* Englewood Cliffs, N.J.: Prentice-Hall.
CAMPION, P. J., BURNS, J. E. and WILLIAMS, A. 1973: *A code of practice for the detailed statement of accuracy.* London: National Physical Laboratory Publication, HMSO.
GREENBERG, L. H. 1975: *Discoveries in physics for scientists and engineers,* second edition. Philadelphia: Saunders.
GREGORY, S. 1973: *Statistical methods and the geographer,* third edition. London: Longmans.
HAMMOND, R. and MCCULLAGH, P. 1974: *Quantitative techniques in geography: an introduction.* Oxford: Oxford University Press.
TOPPING, J. 1972: *Errors of observation and their treatment.* London: Chapman and Hall.

References

ACKOFF, R. L. 1962: *Scientific method: optimising applied research decisions*. New York: Wiley.

ANDERSON, C. B., FORD, P. C. and KENNEDY, J. H. 1973: *Chemistry: principles and applications*. Lexington, Mass: D. C. Heath and Co.

ANDERTON, P. and BIGG, P. H. 1967: *Changing to the metric system*, second edition. London: HMSO.

BAGNOLD, R. A. 1954: *The physics of blown sand and desert dunes*, reprinted edition. London: Methuen.

BARRY, R. G. and CHORLEY, R. J. 1976: *Atmosphere, weather and climate*, third edition. London: Methuen.

BASCOMB, C. L. 1974: Physical and chemical analyses of < 2 mm samples. In Avery, B. W. and Bascomb, C. L., editors, Soil survey laboratory methods, *Soil Survey Technical Monograph* 6, 14–41, Harpenden: Soil Survey of England and Wales.

BIBBY, J. S. and MACKNEY, D. 1969: Land use capability classification *Soil Survey Technical Monograph* 1, Harpenden: Soil Survey of England and Wales.

BIRKELAND, P. W. 1974: *Pedology, weathering, and geomorphological research*. New York: Oxford University Press.

BLOOM, A. L. 1969: *The surface of the earth*. Englewood Cliffs, N.J.: Prentice-Hall.

BOWEN, N. L. 1922: The reaction principle in petrogenesis. *Journal of Geology* **30**, 177–98.

BRADDICK, H. J. J. 1963: *The physics of experimental method*, second edition. London: Chapman and Hall.

BRADY, N. C. 1974: *The nature and properties of soils*, eighth edition. New York: Macmillan.

BRAGG, G. M. 1974: *Principles of experimentation and measurement*. Englewood Cliffs, N.J.: Prentice-Hall.

BRINDLEY, G. W. and MACEWAN, D. N. C. 1953: Structural aspects of the mineralogy of clays and related silicates. In Green, A. T. and Stewart, G. H., editors, *Ceramics: a symposium* 15–59, Stoke-on-Trent: British Ceramic Society.

BRITISH STANDARDS INSTITUTION 1972: *The use of SI units*, PD5686, London.

BROWN, G. 1974: The agricultural significance of clays. In Mackney, D., editor, Soil type and land capability, *Soil Survey Technical Monograph* 6, 27–42, Harpenden: Soil Survey of England and Wales.

BYERS, H. R. 1965: *Elements of cloud physics*. Chicago: University of Chicago Press.

CAMPION, P. J., BURNS, J. E. and WILLIAMS, A. 1973: *A code of practice for the detailed statement of accuracy*. London: National Physical Laboratory Publication, HMSO.

CAPPER, P. L. and CASSIE, W. F. 1963: *The mechanics of engineering soils*, fourth edition. London: Spon.

CAPPER, P. L., CASSIE, W. F. and GEDDES, J. D. 1971: *Problems in engineering soils*, SI edition. London: Spon.

CARSON, M. A. 1971: *The mechanics of erosion*. London: Pion.

CARSON, M. A. and KIRKBY, M. J. 1972: *Hillslope form and process*. Cambridge: Cambridge University Press.

CARSON, M. A. and PETLEY, D. J. 1970: The existence of threshold hillslopes in the denudation of the landscape. *Transactions of the Institute of British Geographers* **49**, 71–95.

CHORLEY, R. J. 1958: Group operator variance in morphometric work with maps. *American Journal of Science* **256**, 208–18.

COLE, F. W. 1970: *Introduction to meteorology.* New York: Wiley.

COUTTS, J. R. H. 1973: Microclimatic conditions. In Glentworth, R. and Muir, J. W., *The soils of the country round Aberdeen, Inverurie and Fraseburgh*, Memoir of the Soil Survey of Great Britain 42–53. Edinburgh: HMSO.

CURTIS, C. D. 1976a: Chemistry of rock weathering: fundamental reactions and controls. In Derbyshire, E., editor, *Geomorphology and climate*, 25–57. London: Wiley.

1976b: Stability of minerals in surface weathering reactions: a general thermochemical approach. *Earth Surface Processes* **1**, 63–70.

DANSGAARD, W., JOHNSEN, S. J. and MØLLER, J. 1969: One thousand centuries of climatic record from Camp Century on the Greenland ice sheet. *Science* **166**, 377–81.

DAWSON, J. A. and UNWIN, D. J. 1976: *Computing for geographers.* Newton Abott: David and Charles.

DIXON, W. J. and MASSEY, F. J. 1957: *Introduction to statistical analysis*, second edition. New York: McGraw-Hill.

DUNCAN, G. 1975: *Physics for biologists.* Oxford: Blackwell.

FITZPATRICK, E. A. 1971: *Pedology: a systematic approach to soil science.* Edinburgh: Oliver and Boyd.

GOLDICH, S. S. 1938: A study in rock weathering. *Journal of Geology* **46**, 17–58.

GREENBERG, L. H. 1975: *Discoveries in physics for scientists and engineers*, second edition. Philadelphia: Saunders.

GREGORY, S. 1973: *Statistical methods and the geographer*, third edition. London: Longmans.

GRIFFITHS, J. C. 1967: *Scientific method in analysis of sediments.* New York: McGraw-Hill.

GYMER, R. G. 1973: *Chemistry: an ecological approach.* New York: Harper and Row.

HAGGETT, P., CLIFF, A. D. and FREY, A. 1977: *Locational analysis in human geography*, second edition, vol. 1. London: Arnold.

HALLIDAY, D. and RESNICK, R. 1966: *Physics*, combined edition. New York: Wiley.

HAMMOND, R. and MCCULLAGH, P. S. 1974: *Quantitative techniques in geography: an introduction.* Oxford: Oxford University Press.

HARRISON, R. 1972: *Atoms and molecules.* Technology Foundation Course Unit 19, The Open University, Bletchley: The Open University Press.

HARVEY, D. 1969: *Explanation in geography.* London: Arnold.

HILLEL, D. 1971: *Soil and water: physical principles and processes.* New York: Academic Press.

HMSO 1973: *The SI system: SI the international system of units.* Prepared jointly by the National Physical Laboratory, UK and the National Bureau of Standards, USA, London: HMSO.

HUDSON, N. 1971: *Soil conservation.* London: Batsford.

ISSS 1963: Soil physics terminology. *Bulletin of the International Society of Soil Science* **23**, 7–10.

JONES, O. T. 1924: The upper Towy drainage system. *Quarterly Journal of the Geological Society of London* **80**, 568–607.

JOYNT, M. I. 1973: Tables of heat capacity of unfrozen soils, Laboratory Technique Report 5, Geotechnical Science, Geography Department, Carleton University, Ottawa, Canada.

JOYNT, M. I. and WILLIAMS, P. J. 1973: The role of ground heat in limiting frost penetration. *Report Vol. 1, Symposium on frost action on roads*, 189–203, Paris: OECD.

JUDGE, A. S. 1973: *The thermal regime of the MacKenzie valley: observations of the natural state*. Report for the Environmental-Social Committee, Northern Pipelines, Task Force on Northern Oil Development, Report 73–38.

JUMIKIS, A. R. 1966: *Thermal soil mechanics*. New Brunswick, N.J.: Rutgers University Press.

——1967: *Introduction to soil mechanics*. Princetown, N. J.: Van Nostrand.

KEENAN, C. W., WOOD, J. H. and KLEINFELTER, D. C. 1976: *General college chemistry*, fifth edition. New York: Harper and Row.

KING, C. A. M. 1966: *Techniques in geomorphology*. London: Arnold.

KOMAR, P. D. 1976: *Beach processes and sedimentation*. Englewood Cliffs, N.J.: Prentice-Hall.

KRUMBEIN, W. C. and GRAYBILL, F. A. 1965: *An introduction to statistical models in geology*. New York: McGraw-Hill.

LEOPOLD, L. B. 1962: Rivers. *American Scientist* **50**, 511–37.

LEOPOLD, L. B. and LANGBEIN, W. B. 1962: The concept of entropy in landscape evolution *US Geological Survey Professional Paper 500–A*, Washington.

LEOPOLD, L. B., WOLMAN, M. G. and MILLER, J. P. 1964: *Fluvial processes in geomorphology*. San Francisco: Freeman.

LINDEMANN, R. L. 1942: The trophic dynamic aspect of ecology. *Ecology* **23**, 399–418.

MCCANCE, R. A. and WIDDOWSON, E. M. 1960: The composition of foods. *Medical Research Council, Special Report Series* 297. London.

MASON, B. J. 1971: *The physics of clouds*, second edition. London: Clarendon Press.

MASSEY, B.S. 1975: *Mechanics of fluids,* third edition. New York: Van Nostrand.

MASTERTON, W. L. and SLOWINSKI, E. J. 1973: *Chemical principles*. Philadelphia: Saunders.

MATHER, P. M. 1976: *Computational methods of multivariate analysis in physical geography*. London: Wiley.

MILLER, A. A. and PARRY, M. 1975: *Everyday meteorology*, second edition. London: Hutchinson.

MONTEITH, J. L. 1973: *Principles of environmental physics* London: Arnold.

MORE, R. J. 1969: The basic hydrological cycle. In Chorley, R. J., editor, *Water, earth and man*, 67–76, London: Methuen.

MORRIS, D. 1973: *The structure and management of ecosystems*. Systems behaviour, module 6, The Open University, Bletchley: The Open University Press.

MORSE, P. M. 1969: *Thermal physics*, second edition. New York: W. A. Benjamin Inc.

MOSLEY, M. P. and ZIMPFER, G. L. 1976: Explanation in geomorphology. *Zeitschrift für Geomorphologie* N. F. **20**, 381–90.

NANSON, G. C. 1974: Bedload and suspended-load transport in a small, steep, mountain stream. *American Journal of Science* **274**, 471–86.

ODUM, E. P. 1971: *Fundamentals of ecology*, third edition. Philadelphia: Saunders.

ODUM, H. T. 1971: *Environment, power and society*. New York: Wiley.

OLLIER, C. D. 1969: *Weathering*. Edinburgh: Oliver and Boyd.

PARRISH, A. 1969: *SI conversion charts for imperial and metric quantities*. London: Iliffe Books.

PAULING, L. 1970: *General chemistry*, third edition. San Francisco: Freeman.

PIERCE, J. B. 1970: *The chemistry of matter*. Boston: Houghton Mifflin.

REYNOLDS, W. C. 1974: *Energy: from nature to man*. New York: McGraw-Hill.

ROGERS, R. R. 1976: *A short course in cloud physics*. Oxford: Pergamon Press.

ROUSE, W. C. and FARHAN, Y. I. 1976: Threshold slopes in south Wales. *Quarterly Journal of Engineering Geology* **9**, 327–38.

ROYAL SOCIETY 1975: *Quantities, units and symbols*. London.

SCHEIDEGGER, A. E. 1970: *Theoretical geomorphology*, second edition. Berlin: Springer-Verlag.

SEARS, F. W. and ZEMANSKY, M. W. 1963: *University physics*, part 1, third edition. Reading, Mass.: Addison-Wesley.

SEARS, F. W., ZEMANSKY, M. W. and YOUNG, H. D. 1974: *College physics*, fourth edition. Reading, Mass.: Addison-Wesley.

SELLERS, W. D. 1965: *Physical climatology*. Chicago: University of Chicago Press.

SIENKO, M. J. and PLANE, R. A. 1974: *Chemical principles and properties*, second edition. New York: McGraw-Hill.

——1976: *Chemistry*, fifth edition. New York: McGraw-Hill.

SLOBODKIN, L. B. 1959: Energetics in *Daphnia pulex* populations. *Ecology* **40**, 232–43.

SMITH, H. G. 1971: *Minerals and the microscope*, fourth edition. London: Murby.

SPARKS, B. W. 1953: Effects of weather on the determination of heights by aneroid barometer. *Geographical Journal* **119**, 73–80.

STAMPER, J. G. and STAMPER, M. A. 1971: *Chemistry for biologists*. London: Allen and Unwin.

STATHAM, I. 1976: A scree slope rockfall model. *Earth Surface Processes* **1**, 43–62.

STRAHLER, A. N. and STRAHLER, A. H. 1973: *Environmental geoscience*. Santa Barbara, Calif.: Hamilton.

——1974: *Introduction to environmental science*. Santa Barbara, Calif.: Hamilton.

STREETER, V. L. and WYLIE, E. B. 1975: *Fluid Mechanics*, sixth edition. New York: McGraw-Hill.

SUMNER, G. N. 1978: *Mathematics for physical geographers*. London: Arnold.

TAYLOR, S. A. and JACKSON, R. D. 1965: Heat transfer. In Black, C. A., editor, *Methods of soil analysis*, Part 1, 349–60, Madison, Wisconsin: American Society for Agronomy.

TERZAGHI, K. 1943: *Theoretical soil mechanics*. New York: Wiley.

TULLEY, D. and THUMM, W. 1971: *College physics with applications to the life sciences*. Menlo Park, Calif.: Cummings Publishing Co.

TOPPING, J. 1972: *Errors of observation and their treatment*, fourth edition. London: Chapman and Hall.

TOWNSEND, W. R. 1973: *An introduction to the scientific study of the soil*, fifth edition. London: Arnold.

UNESCO 1970: Combined heat, ice and water balances at selected glacier basins. *Technical papers in Hydrology* 5. Paris.

VENNARD, J. K. and STREET, R. L. 1975: *Elementary fluid mechanics*, fifth edition. New York: Wiley.

WATTS, D. 1971: *Principles of biogeography*. London: McGraw-Hill.

WEAST, R. C. (editor) 1974: *Handbook of chemistry and physics*, fifty-fifth edition. Cleveland, Ohio: Chemical Rubber Co.

WEAVER, G. 1972: *Energy in chemical reactions*. Technology Foundation Course Unit 24, The Open University, Bletchley: The Open University Press.

WHALLEY, W. B. 1976: *Properties of materials and geomorphological explanation*. London: Oxford University Press.

WHITTLE, R. M. and YARWOOD, J. 1973: *Experimental physics for students*. London: Chapman and Hall.

WILSON, A. G. and KIRKBY, M. J. 1975: *Mathematics for geographers and planners*. London: Oxford University Press.

YANG, C. T. 1972: Unit stream power and sediment transport. *Proceedings of the American Society of Civil Engineers: Journal of the Hydraulics Division* **98**, part HY10, 1805–1826.

YOUNG, A. 1972: *Slopes*. Edinburgh: Oliver and Boyd.

Index

Information about subjects in tables or diagrams is indicated by page numbers in italics.